East European Poetry: An Anthology

CONTEMPORARY EAST EUROPEAN POETRY

AN ANTHOLOGY

Edited by Emery George

30036001644457

OXFORD UNIVERSITY PRESS
New York   Oxford

Oxford University Press

Oxford   New York   Toronto
Delhi   Bombay   Calcutta   Madras   Karachi
Kuala Lumpur   Singapore   Hong Kong   Tokyo
Nairobi   Dar es Salaam   Cape Town
Melbourne   Auckland   Madrid

and associated companies in
Berlin   Ibadan

First published in 1983 by Ardis Publishers
2901 Heatherway, Ann Arbor, Michigan 48104

First issued as an Oxford University Press paperback, 1993

Oxford is a registered trademark of Oxford University Press

Library of Congress Cataloging-in-Publication Data
Contemporary East European poetry : an anthology / edited by Emery
   George.
   p. cm.   Originally published: Ann Arbor : Ardis, c1983.
   ISBN 0-19-508635-X. — ISBN 0-19-508636-8 (pbk.)
   1. East European poetry—20th century—Translations into English.
   I. George, Emery Edward, 1933–   .
   PN849.E92C66 1993   93-28192   891.8—dc20

2 4 6 8 10 9 7 5 3 1

Printed in the United States of America

To the memory of
Carl R. Proffer
(1938–1984)

PREFACE

Contemporary East European Poetry first appeared with Ardis Publishers in June 1983, after well over three years of preparation; its gestation period was the late 1970s and early 1980s. I recall submitting the manuscript in August 1981, just prior to my own departure for Hungary as an IREX East European Research Fellow. It was, as always, the best and the worst of times. It was the time of Brezhnev and of Reagan, of the "evil empire" and of the ill-starred Korean Air Lines Flight 007. The cold war was very much on, and there seemed to be no light at the end of that nightmarish tunnel—with the one significant exception of arrangements enabling scholars on either side to conduct research in one another's countries.

A great deal has changed in the decade since. Bad has changed to both better and worse; the mutual terror of the cold war has given way to a new age of anxiety. If the threat of Communism in its worst forms is over, it is being replaced by another threat: renewed nationalism and xenophobia, at times of genocidal intensity. Yet ours is also a good time for the humanities and arts, especially poetry. Poetry, thank heaven, is like a bad weed: there is not much anyone can do to kill it. The best poets, now as during both world wars, will go on writing, almost no matter what their circumstances. This truth needs a careful renewed look beyond the clichés about poetries of "dissent." All poetry is the poetry of dissent. It is born, among other instigations, of the individual artist's disagreement with his or her own reluctances. Poetry is an act of overcoming; internally, it is certainly a political act. In looking ahead, poets will do well to heed this truth if they are to go on producing while deprived of the stimulations that have proven to be such a healthy byproduct of repression.

This new edition of *Contemporary East European Poetry* is, I believe, as significant and timely an undertaking as its predecessor. It presents a prime opportunity to show both continuity and contrast in one of the world's great bodies of poetry. While the bulk of the volume preserves the texts first presented ten years ago, a closing section takes account of a new generation that in the interim has come into its own. The carefully chosen poems in the final pages update eight major poetic cultures: Polish, German, Czech and Slovak, Hungarian, Romanian, Yugoslav (Serbian and Macedonian), Bulgarian, and Yiddish. They range from works by major poets of the 1930s not included earlier, to texts by poets born between 1943 and 1972. I am pleased to include poems by Ernest Bryll (b. 1935), Volker Braun (b. 1939), Dezső Tandori (b. 1938), and Anghel Dumbrăveanu (b. 1933), four vastly different writers, and fresh voices regardless of chronological age. In some cases what should count is not date of birth but time of composition. "In a Fever," by

Ernest Bryll, bears the date 6 April 1986, and there can be no doubt that the event depicted corresponds to real events that transpired under Solidarity. Along with Stanisław Barańczak's "The Three Magi," Bryll's poem is also an example of successful political poetry, a subgenre as rare as it is instructive. We seldom reflect that it is as ancient a form, and as difficult to cultivate, as the sonnet.

Let me pause at the Yiddish contribution to our volume. While Yiddish is fast joining such "dead" languages as classical Greek and Latin, the poetry continues to be an important phenomenon in East European letters. These poems are a major discovery and a study in contrast. Readers looking for lyric variety need but juxtapose Malka Heifetz-Tussman's mythical imagery with Melech Ravitch's hard-won topicality, H. Leyvik's controlled lyricism with Berysh Vaynshteyn's lyric-epic diction. And this spectrum of energies and textures is present in the earlier section as well.

In the update section the younger poets are, of course, the majority; twenty-three of the thirty-two were born in or since 1945, among them some gifted women. I am delighted to be able to represent such vital new talents as Helga Marie Heinze (b. 1946), Sylva Fischerová (b. 1963), Zsuzsa Rakovszky (b. 1950), Grete Tartler (b. 1948), Novica Tadić (b. 1949), and Georgi Borisov (b. 1950). Distribution and weighting once again presented problems of their own; while Slovak poetry is present, Croatian and Slovenian are not, to my regret. Even so, the total picture looks encouraging; East European poetry is being vigorously translated into English these days. Faced with an abundance of good material, I often found choosing difficult. As before, quality and not ideological slant was the criterion of inclusion. Here, the poem comes first.

Once again I would like to express my warm thanks to individuals and institutions who have furthered my work: Robert Austerlitz, Marvin Bielawski, Bogdana Carpenter, John Carpenter, Eric Han, Seymour Mayne, Adriana Popescu, Marianne Sághy, Howard Schwartz, Jonas Zdanys; Columbia University Library, Princeton University Library, and the YIVO Institute for Jewish Research. To my wife, Mary, I am grateful for the solution of more than one knotty bibliographic problem. I thank all our contributors, as well as publishers and other agencies who have given their gracious permission to print or reprint these translations, as specified in the updating section in the Acknowledgments. To my editors at Oxford University Press, Ellen I. Chodosh and Linda D. Robbins, as well as to Peter Bakel, I am indebted for encouragement, advice, and assistance in more ways than I could possibly list.

Princeton, New Jersey   —Emery George
June 1993

CONTENTS

Like a Black Hedgehog † In Spring during Spring

xi

xiii

xiv

xv

YUGOSLAV

YIDDISH

THE UPDATE SECTION

POLISH

GERMAN

YIDDISH

ACKNOWLEDGMENTS

Every effort has been made to trace owners of copyright. Additional information will appear in future editions of this book.

For permission to print or reprint English translations of poems grateful acknowledgment is hereby made to agencies, individuals, and publishers named following the name of each poet.

BALTIC

Estonian

BETTI ALVER: "Tähetund," "Kild" from *Tähetund*, © 1966 Eesti Raamat, Tallinn. "Tuulelapsed," "Raudsed närvid" from *Eluhelbed*, © 1971 Eesti Raamat, Tallinn. Translations "Stellar Hour," "Fragment," "Children of the Wind," "Iron Nerves" © 1981 Ivar Ivask and printed with his permission.

IVAR GRÜNTHAL: "Ma murdsin esimese käsu teise osa," "Käib vahetpidamata tähelangus." "Nüüd pulbriks tehtud on kõik suured mehed," "Nii viibib veel endaga kõnes" from *Must pühapäev*, © 1954 Vaba Eesti, Stockholm. Translations "I Broke the Second Part of the First Commandment," "A Constant Starburst Showers Us," "All Great Men Were Ground to Powder," "And Thus a Monologue Keeps Flowing" © 1981 Ivar Ivask and printed with his permission.

IVAR IVASK: "Me läheme tagasi me jääme siia," "Lehestik lehestik sügistuule tähestik," "Ja siis ma jutustasin Oklahoma punasest maast," "Olen taevaste lambakarjus" from *Elukogu*, © 1978 Eesti Kirjanike Kooperatiiv, Lund. Translations "We Shall Return We Shall Stay Here," "Foliage, foliage," "And Then I Told about the Red Soil of Oklahoma," "I Am the Shepherd of the Heavenly Flock" © 1981 by the author and printed with his permission.

BERNARD KANGRO: "Vanadus," "Hilised lilled, tuul, meri, liiv ja kalad," "Puu kõnnivad kaugemale" from *Minu nägu*, © 1970 Eesti Kirjanike Kooperatiiv, Lund. Translations "Old Age," "Late Flowers, Wind, Sea, Sand and Fish," "The Trees Walk Farther Away" © 1981 Ivar Ivask and used with his permission.

JAAN KAPLINSKI: "Valge ristikhein ei küsi küll midagi" from *Tolmust ja värvidest*, © 1967 Perioodika, Tallinn. "Sain aru sain aru," "Kevadel kevades" from *Ma vaastasin päikese aknasse*, © 1976 Eesti Raamat, Tallinn. "Suure musta siilina" from *Uute kivide kasvamine*, © 1977 Eesti Raamat, Tallinn. Translation "White Clover Will Not Ask Anything" © 1981 Ilse Lehiste and used with her permission. "I Understood I Understood" © 1981 Hellar Grabbi and used with his permission. "In Spring during Spring," "Like a Black Hedgehog" © 1981 Ivar Ivask and used with his permission.

KALJU LEPIK: "Meri," "Las ma hoian veel su käsi," "Needmine," "Mulla keelt keegi ei kõnele" from *Kogutud luuletused*, © 1980 Eesti Kirjanike Kooperatiiv, Lund. "The Sea," "Let Me Hold Your Hand," "Curse," "Nobody Talks Earth's Language" © 1981 Ivar Ivask and used with his permission.

UKU MASING: "Udu ainult on kindel" from *Neemed vihmade lahte*, © 1959 Vaba Eesti, Stockholm. "Pollale," "Igale usklikule" from *Udu Toonela jõelt*, © 1974 Maarjamaa, Rome. Translations "Only the Mists Are Real," "For Polla," "To Every Believer" © 1981 Ivar Ivask and used with his permission.

ALEKSIS RANNIT: "Nii näen sind ikka," "Wiiralt joonistab Chartres'is," "Boris Vilde" from *Suletud avarust*, © 1956 Eesti Kirjanike Kooperatiiv, Lund. "Jaan Oks" from *Kuiv hiilgus*, © 1963 Eesti Kirjanike Kooperatiiv, Lund. "So I See You Still," trans. Henry Lyman. First published (as "So I See Thee Still") in *New Directions in Prose and Poetry*, ed. J. Laughlin, © 1972 by New Directions Publishing Corporation. Reprinted with the permission of New Directions Publishing Corporation. Also in Aleksis Rannit, *Cantus firmus*, trans. Henry Lyman, published by The Elizabeth Press, New Rochelle, New York. © 1978 Aleksis Rannit. Reprinted by permissin of Aleksis Rannit and of the translator. "Wiiralt Drawing in Chartres" © 1981 Emery George and used with his permission. "Boris Vilde" and "Jaan Oks," trans. Henry Lyman. "Jaan Oks" published in Aleksis Rannit, *Line*, © 1970 Adolf Hürlimann, Zürich. "Jaan Oks" and "Boris Vilde" published in *Signum et verbum* by the Elizabeth Press, New Rochelle, New York. © 1981 Aleksis Rannit. Used by permission of Aleksis Rannit and of the translator.

PAUL-EERIK RUMMO: "Jälle jälle," "Taevas kummargil üle maa," "Siin oled kasvanud," "Maailm mu hinge" from *Luulet 1960-1967*, © 1968 Eesti Raamat, Tallinn. Translations "Again Again," "The Sky Bends over the Earth," "Here You Grew Up," "The World Did Not Force Itself into My Soul" © 1981 Ivar Ivask and used with his permission.

MARIE UNDER: "Hommikurõõm," "Üksi merega," "Ja langes üks täht" from *Kogutud luuletused*, © 1958 Vaba Eesti,

Stockholm. Translations "Morning Joy," "Alone with the Sea," "And a Star Fell" © 1981 Ivar Ivask and printed with his permission.

JUHAN VIIDING: "George Marrow 1011. uni" from *Detsember,* © 1971 Eesti Raamat, Tallinn. "Olen pärisori," "Ma pühitsesin seda päeva märtsis," "Kõned sisemaal" from *Käekäik,* © 1973 Perioodika, Tallinn. Translations "George Marrow's 1011th Dream," "I Am a Serf," "I Celebrated This Day in March," "Speeches in the Interior" © 1981 Ivar Ivask and printed with his permission.

Latvian

VIZMA BELŠEVICA: "To Be Roots," "Don't Be Oversure," "That Was," "Willow-Catkins" from *Jura deg,* © 1966 Liesma, Riga, and from *Gada gredzeni,* © 1969 Liesma, Riga. Translations © 1981 Inara Cedrins and Valters Nollendorfs and used with their permission.

MARIS ČAKLAIS: "Give Me Your Silver Knife," "Dedication," "Woman in Mourning," original texts appeared in *Sastrgumstun* © 1974 Liesma, Riga. Translations © 1981 Inara Cedrins and printed with her permission.

ANDREJS EGLITIS: "Winter Has Already Shriveled," "In a Naples Bar," "Black Doves," "I See How" from the unpublished cycle "Otranto." Translations © 1981 Inara Cedrins and used with her permission.

ASTRIDE IVASKA: "For My Godmother," "Autumn in the Cascade Mountains," "K.H.," "To the Memory of a Poet" from *At the Fallow's Edge,* by Astrid Ivask, ed. and trans. Inara Cedrins, published by Inklings 3, Santa Barbara, 1981. © 1980 Astrid Ivask; translations © 1980 Inara Cedrins. The original texts first appeared in *Solis silos,* © 1973 Daugava, Sweden, and in *Ziemas tiesa,* © 1968 The River Hill Press, Shippenville, Pennsylvania. Translations reprinted with the permission of the author and the translator. "That Which Has Remained Unlived" © 1981 Inara Cedrins and used with her permission.

AINA KRAUJIETE: "City Girls," "In Chalk Rooms," "Vagabonds Arising," original texts first published in *No aizpriktas Paradizes,* © 1966 The River Hill Press, Shippenville, Pennsylvania. Translations © 1981 Inara Cedrins and printed with her permission.

GUNARS SALINŠ: "A Bomb Victim in the Cathedral Speaks to a Child," "Song," © 1981 Gunars Salinš. Translation of "A Bomb Victim" © 1981 Baiba Kaugara; translation of "Song" © 1981 Laris Salinš. Poems used with the permission of the poet and of the translators.

OLAFS STUMBRS: "Cigars," "Ode to a Country Childhood," "A Light, Late Song for Solo Voice without Piano," "A Married Romantic" © 1981 Olafs Stumbrs. "Ode to a Country Childhood" and "A Light, Late Song" also translated by Olafs Stumbrs. Translations © 1981 Olafs Stumbrs. Poems printed by permission of Olafs Stumbrs.

OJARS VACIETIS: "Burning Leaves," "Sunday," "Horse Dream," "Of Nights," original texts published in *Gamma,* © 1976 Liesma, Riga. Translations © 1981 Inara Cedrins and used with her permission.

IMANTS ZIEDONIS: "How the Candle Burns," "Inevitability," "That Is Her Memorial," "That Was a Beautiful Summer," original texts first appeared in *Ka svece deg,* © 1971 Liesma, Riga. Translations © 1981 Inara Cedrins and used with her permission.

Lithuanian

VYTAUTAS BLOŽE: "Lapai pagelto" appeared in the anthology *Poezijos Pavasaris,* © 1970 Vaga, Vilnius. Translation "The Leaves Have Yellowed" © 1981 Irene Pogożelskyte Suboczewski and used with her permission.

JANYNA DEGUTYTE: "Džiazo ritme," "Stiklinis etiudas" from *Pilnatis,* © 1967 Vaga, Vilnius. Translations "In Jazz Rhythms," "Etude in Glass" © 1981 M.G. Slavenas. Printed by permission of M.G. Slavenas.

SIGITAS GEDA: "I Walked Out into Lithuania," "Repentance: The Devil's Blossom," "God's Family," "Sebastian's Lament, 1943," "Steep Eyes of Wooden Gods" from *Selected Post-War Lithuanian Poetry,* ed. and trans. Jonas Zdanys. © 1978 Manyland Books, Woodhaven, New York. Reprinted by permission of Manyland Books, Inc. and of the translator.

JUSTINAS MARCINKEVIČIUS: "Peizažas su regejimu" from *Poezijos Pavasaris,* ed. Vytautas Rudokas, © 1970 Vaga, Vilnius. "Nemunas" from the anthology *Poezijos Pavasaris,* ed. Antanas Drilinga, © 1976 Vaga, Vilnius. Translation "Landscape with Apparition," translated by Irene Pogożelskyte Suboczewski, first appeared in *Mr. Cogito,* Vol. 5, no. 2, 1981. Reprinted by permission. "Nemunas" © 1981 Irene Pogożelskyte Suboczewski and used with her permission. "Father's Winter" reprinted from *Selected Post-War Lithuanian Poetry,* ed. and trans. Jonas Zdanys. © 1978 Manyland Books, Woodhaven, New York. Reprinted with the permission of the publishers and of the translator.

EUGENIJUS MATUZEVIČIUS: "Ugnies rašmenys" from the anthology *Poezijos Pavasaris,* ed. Albinas Bernotas, © 1975 Vaga, Vilnius. "The Writing in the Fire" © 1981 Irene Pogożelskyte Suboczewski and used with her permission.

xxvii

HUNGARIAN

"The Jaguar Is Getting Ready," "After All" © 1981 Juliette Victor-Rood and printed with her permission. Permission to print translations from the estate of Anna Hajnal granted by Erzsébet Holczer, "Artisjus," and Magyar P.E.N.

GYULA ILLYÉS: "Regényrészlet" from *Haza a magasban*, published by Szépirodalmi, Budapest. © 1972 Gyula Illyés. "Meggyfák" from *Teremteni*, published by Szépirodalmi, Budapest. © 1973 Gyula Illyés. "Tág tél" from *Minden lehet*, published by Szépirodalmi, Budapest. © 1973 Gyula Illyés. Translations "Part of a Novel," "Sour-Cherry Trees," "Spacious Winter" © 1981 Emery George and printed with his permission. "Tökéletes alkony" from *Minden lehet*, published by Szépirodalmi, Budapest. © 1973 Gyula Illyés. Translation "Deep Dusk" © 1981 Nicholas Kolumban and printed with his permission. Permission to print translations granted by Gyula Illyés.

MÁRTON KALÁSZ: "Fiatal lány" from *Rapszódiáink évada*, published by Magvető, Budapest. © 1963 Márton Kalász. "Örökség" from *Ünnep előtt*, published by Szépirodalmi, Budapest. © 1961 Márton Kalász. "Himnusz" from *Változatok a reményre*, published by Magvető, Budapest. © 1967 Márton Kalász. "Impromptu," "Az idő nem grammatika," "Mindenütt még ily falba ütközöm" from *Az imádkozó sáska*, published by Magvető, Budapest. © 1980 Márton Kalász. Translations "Maiden," "Legacy," "Hymn," "Impromptu," "Time's No Grammar," "Up against that Wall Everywhere" © 1981 Jascha Kessler and printed with his permission. Translations also printed in magazines as follows: "Maiden" in *The New Hungarian Quarterly* and in *The Spirit that Moves Us;* "Legacy" in *The American Poetry Review;* "Hymn" in *California Quarterly*. Translations of poems by Márton Kalász printed by permission of the poet.

ÁGNES NEMES NAGY: "Fenyő," "A lovak és az angyalok," "Fügefák," "Egy költöhöz," "Vihar," "Ekhnáton éjszakája" from *A lovak és az angyalok*. *Válogatott versek*, published by Magvető, Budapest. © 1969 Ágnes Nemes Nagy. Translations "Pinetree," "The Horses and the Angels," "Figtrees," "To a Poet" from Ágnes Nemes Nagy, *Selected Poems*, trans. and ed. Bruce Berlind, published by Iowa Translations, International Writing Program, The University of Iowa. © 1980 The University of Iowa. International Writing Program. Reprinted by permission of the International Writing Program and of Bruce Berlind. Translations "Ikhnaton's Night," "Storm" appeared in *The Penguin Book of Women Poets*, ed. Carol Cosman, Joan Keefe, and Kathleen Weaver, Penguin, New York, 1978. © 1981 Laura Schiff and printed with her permission. Poems by Ágnes Nemes Nagy in English translation printed by permission of Ágnes Nemes Nagy.

OTTÓ ORBÁN: "Békeoszlop," "Cigánykarácsony," "Az az erő," "Nyár a tavon," "Rekviem" from *Szegénynek lenni*, published by Magvető, Budapest. © 1974 Ottó Orbán. "Azok a bizonyos évek" from *Az alvó vulkán*, published by Magvető, Budapest. © 1981 Ottó Orbán. Translations "Peace Pillar," "Gypsy Christmas," "That Strength." © 1981 Emery George and printed with his permission. Translations "A Summer on the Lake," "Requiem," "Certain Years" © 1981 Jascha Kessler and printed with his permission. "A Summer on the Lake" previously published in *The Massachusetts Review*, Summer 1977; "Certain Years" appeared in *Midstream*, February 1980. "Hóesés Bostonban" from *A visszacsavart láng*, published by Magvető, Budapest. © 1979 Ottó Orbán. Translation "Snowfall in Boston" © 1981 Tímea K. Szell and printed with her permission. Translations of poems by Ottó Orbán published with the permission of Ottó Orbán.

JÁNOS PILINSZKY: "Parafrázis," "Aranykori töredék," "Utószó" from *Nagyvárosi ikonok*, published by Szépirodalmi, Budapest. © 1971 the estate of János Pilinszky ("Artisjus"). "A mélypont ünnepélye" from *Szálkák*, published by Szépirodalmi, Budapest. © 1972 the estate of János Pilinszky ("Artisjus"). Translations "Paraphrase," "Fragment from the Golden Age," "Afterword," "Celebration of Nadir" © 1981 Emery George and printed with his permission. "In memoriam F.M. Dosztojevszkij," "Sztavrogin elköszön," "Sztavrogin visszatér" from *Végkifejlet*, published by Szépirodalmi, Budapest. ~ 1974 the estate of János Pilinszky ("Artisjus"). Also published in *Kráter*, Szépirodalmi, Budapest. © 1976 the estate of János Pilinszky ("Artisjus"). Translations "In memoriam F.M. Dostoevsky," "Stavrogin's Farewell," "Stavrogin's Return" © 1981 Jascha Kessler and printed with his permission. Poems by János Pilinszky in English translation printed by permission of "Artisjus," Budapest.

JUDIT TÓTH: "Erdő," "Külváros, délután," "Zúgás," "Maszkok," "Február," "Két város" from *Két város*, published by Szépirodalmi, Budapest. © 1972 Judit Tóth. Translations "Forest," "Outskirts, Afternoon," "Murmuring," "Masks," "February," "Two Cities" © 1981 Emery George and printed with his permission. Translations of poems by Judit Tóth printed by permission of "Artisjus."

ISTVÁN VAS: "Katakombák," "Mi maradt?" from *Rapszódia a hűségről*, published by Szépirodalmi, Budapest. © 1977 István Vas. Translations "Catacombs," "What Is Left?" previously appeared in *Voices within the Ark: The Modern Jewish Poets*, ed. Howard Schwartz and Anthony Rudolf, published by Avon, New York, 1980. © 1981 Emery George and published with his permission. "Csak azt az egyet," "Egy táncdobra" from *Önarckép a hetvenes évekből*, published by Szépirodalmi, Budapest. © 1974 István Vas. Translations "Just This," "Tambour" appeared in *Contemporary Quarterly*, Vol. 2, no. 4, 1978, and in *Blue Unicorn*, Vol, 1, no. 2, 1977, respectively. © 1981 Jascha Kessler and printed by his permission. Translations of poems by István Vas printed by permission of István Vas.

SÁNDOR WEÖRES: "Shakespeare és Velázquez szelleméhez," "Kínai templom," "A manó" from *A hallgatás tornya. Harminc év verseiből*, published by Szépirodalmi, Budapest. © 1956 Sándor Weöres. "Merülő Saturnus," "Ecce homo" from *111 vers*, published by Szépirodalmi, Budapest. © 1974 Sándor Weöres. Translations "To the Spirits of Shakespeare and Velázquez," "Chinese Temple," "Elf," "Saturn Sinking," "Ecce homo" © 1981 Emery George and printed by his permission. Translations of poems by Sándor Weöres printed by permission of Sándor Weöres.

ROMANIAN:

Translations in this section printed by permission secured through the kind assistance of the Academia de Stiinte Sociale si Politice (Academy of Social and Political Sciences) and of the Uniunea Scriitorilor din Republica Socialista România (Union of Writers of the RSR), Bucharest.

IOAN ALEXANDRU: "Omul," "Rugăciune" from *Infernul discutabil,* © 1966. "Miserere nobis," "Corborind" from *Vina,* © 1967. "Man" © 1981 Andrei Bantas and Thomas A. Perry and printed with their permission. "Prayer" translated by Dan Dutescu and reprinted from *Romanian Poets of Our Time,* © 1974 Univers Publishing House, Bucharest. Printed with the permission of Univers Publishing House and of Dan Dutescu. "Miserere nobis" appeared in *PN Review,* Vol. 6, no. 2, translated by Peter Jay and Virgil Nemoianu. © 1981 Peter Jay and reprinted with his permission. "Descending" reprinted from *Modern Romanian Poetry: An Anthology,* trans. and ed. Donald Eulert and Stefan Avadanei, © 1973 Donald Eulert and Editura Junimea, Iasi. Reprinted with the permission of Donald Eulert. Translations of poems by Ioan Alexandru printed with the permission of Ioan Alexandru.

TUDOR ARGHEZI: "Testament," "Psalm," "În golf" from *Scrieri,* Vol. 1, © 1962 Editura pentru literatura, Bucharest. "Testament" translated by Andrei Bantas and Thomas A. Perry. © 1981 Andrei Bantas and Thomas A. Perry and printed by their permission. "Psalm" translated by Peter Jay and Virgil Nemoianu and reprinted from *Agenda,* Vol. 12, no. 3, 1974. Translation © 1981 Peter Jay and used by his permission. "Gulf" translated by Stavros Deligiorgis and originally published in *Modern Poetry in Translation,* nos. 37/38, Winter 1979, p. 33. Reprinted by permission of the editor, Daniel Weissbort. Poems by Tudor Arghezi in English translation printed by kind permission of the Academia de Stiinte Sociale si Politice, Bucharest.

ION BARBU: "Din ceas, dedus," "Dioptrie," "Timbru," "Oul dogmatic" from *Joc secund / Jeu second,* with trans. into French by Yvonne Stratt, © 1973 Editura Eminescu, Bucharest. "Counter Play" © 1981 Thomas A. Perry and George Preda and printed with their permission. "Dioptric," translated by Thomas A. Perry, published in *Paintbrush, Vol. 4,* nos. 7 and 8, Spring and Autumn 1977. © 1977 Ishtar Press, Inc., and printed with their permission. "Timbre" translated by Dan Dutescu, reprinted from *The Romanian Review.* © 1981 Dan Dutescu and printed by his permission. "The Dogmatic Egg" reprinted from *Modern Romanian Poetry: An Anthology,* trans. and ed. Donald Eulert and Stefan Avădanei, © 1973 Donald Eulert and Editura Junimea, Iasi. Reprinted with the permission of Donald Eulert. Translations of poems by Ion Barbu printed by permission of the Academia de Stiinte Sociale si Politice, Bucharest.

LUCIAN BLAGA: "În munti," "Veac," "Sufletul satului," "Heraclit lînga lac," "Anno Domini" from *Poemele luminii / Poems of Light,* A Romanian-English Bilingual Edition, trans. and ed. Don Eulert, Stefan Avădanei, Mihail Bogdan, © 1975 Editura Minerva, Bucharest. "In the Mountains," translated by Peter Jay and Virgil Nemoianu, published in *PN Review,* Vol. 6, no. 2. Translation © 1981 Peter Jay and printed by his permission. "Twentieth Century" © 1981 Mihail Bogdan and printed by his permission. "The Soul of the Village" reprinted from *Poemele luminii / Poems of Light,* A Romanian-English Bilingual Edition, trans. and ed. Don Eulert, Stefan Avădanei, Mihail Bogdan, © 1975 Editura Minerva, Bucharest. Reprinted by permission of Mihail Bogdan. "Heraclitus by the Lake" translated by Peter Jay and Virgil Nemoianu. © 1981 Peter Jay and printed by his permission. "Anno Domini" translated by Peter Jay and Virgil Nemoianu. © 1981 Peter Jay and printed by his permission. All translations of poems by Lucian Blaga printed by the kind permission of the Academia de Stiinte Sociale si Politice, Bucharest.

ANA BLANDIANA: "Dintr-un sat," "Cîntec," "Dans în ploaie" from *Poeme,* in the series Cele mai frumoase poezii, © 1978 Editura Albatros, Bucharest. Translations by Irina Livezeanu, "From a Village," "Song," "Dance in the Rain" printed by permission of Editura Albatros and of Irina Livezeanu. "Legături" from *Poeme,* in the series Cele mai frumoase poezii, © 1978 Editura Albatros, Bucharest. Translation "Links" by Michael Impey reprinted from *Shantih,* Vol. 3, no. 4, 1977. © 1981 Michael Impey and printed by his permission. Translations of poems by Ana Blandiana printed by permission of Ana Blandiana.

ION CARAION: "Memorie," "Despodobire," "Podeuna in martie," "Învaluitorul ecou," "Mătase măruntă," "Iarnă," "La lumina crengilor afara" from *Necunoscutul ferestrelor,* © 1969 Editura pentru literatură, Bucharest. "Hipocamp arhaic" from *Cîrtita si aproapele,* © 1970 Editura Eminescu, Bucharest. "Memory," "Deornamentation," "Always, in March," "The Enveloping Echo," "By the Light of the Branches Outside," "Bits of Silk," "Winter," "Archaic Hippocamp" published in *Ion Caraion: Poems,* translated by Marguerite Dorian and Elliott B. Urdang, © 1981 Ohio University Press. Reprinted by permission of Ion Caraion, Ohio University Press, and of the translators.

NINA CASSIAN: "Pubertate," "Ispita," "Apoi" from *Disciplina harfei,* © 1964. "Tinarul vampir" from *Ambitus,* © 1969. "Singele" from *Chronophagia,* © 1970. Translation "Puberty" by Brian Swann and Michael Impey and reprinted from *Shantih,* Vol. 3, no. 4, 1977 © 1981 Brian Swann and reprinted with his permission. "Temptation" translated by Brenda Walker and Andrea Deletant. © Brenda Walker and Andrea Deletant. "Then" translated by Laura Schiff and Virgil Nemoianu, published in *Lady of Miracles,* trans. Laura Schiff and Virgil Nemoianu, © 1981 Cloud Marauder Press, Berkeley. Reprinted by permission of the publisher and of the translators. "The Young Vampire," "The Blood" translated by Marguerite Dorian and Elliott B. Urdang. © 1981 Marguerite Dorian and Elliott B. Urdang and printed by their permission. Translations of poems by Nina Cassian printed by permission of Nina Cassian.

MIRIAM ULINOVER: "Havdolah Wine," "In the Courtyard" translated by Seth L. Wolitz. © 1979 Seth L. Wolitz and printed with his permission.

MOSHE YUNGMAN: "Melons" translated by Gabriel Preil and Howard Schwartz and reprinted from *Lyrics and Laments: Selected Translations from Hebrew and Yiddish*, by Howard Schwartz, BkMk Press, 1979. © 1979 Gabriel Preil and Howard Schwartz and reprinted with their permission.

Note: The poems by Alquit-Blum, Asya, Glanz-Leyeles, Halpern, Korn, Kulbak, Leib, Manger, Segal, Teller, Ulinover, Yungman, and the first five poems named by Sutskever also appeared in *Voices within the Ark: The Modern Jewish Poets*, ed. Howard Schwartz and Anthony Rudolf, Avon Books, New York, 1980. © 1980 Howard Schwartz and Anthony Rudolf. Reprinted with the permission of the editors.

ADDITIONS TO THE ORIGINAL ACKNOWLEDGMENTS

Note: All permissions have been renewed. Information under this heading acknowledges copyright in, and permission to print or reprint, translations included in selections of poetry published since this anthology first appeared. Care has been taken to identify translations by the translators who originally contributed their work to our volume.

LATVIAN

VIZMA BELŠEVICA: Translations "Don't Be Oversure," "That Was a Polite Fish," translated by Inara Cedrins, as in *Contemporary Latvian Poetry*, edited by Inara Cedrins. Iowa Translations series. Copyright © 1984 by The University of Iowa. Used by kind permission of University of Iowa Press and Inara Cedrins.

MĀRIS ČAKLAIS: Translations "Give Me Your Knife," "Dedication," translated by Inara Cedrins, as in *Contemporary Latvian Poetry*, edited by Inara Cedrins. Iowa Translations series. Copyright © 1984 by The University of Iowa. Used by kind permission of University of Iowa Press and Inara Cedrins.

ASTRIDE IVASKA: Translation "K.H.," translated by Inara Cedrins, as in *Contemporary Latvian Poetry*, edited by Inara Cedrins. Iowa Translations series. Copyright © 1984 by The University of Iowa. Printed by kind permission of University of Iowa Press and Inara Cedrins.

AINA KRAUJIETE: Translations "Vagabond's Arising," "City Girls," "In Chalk Rooms," translated by Inara Cedrins, as in *Contemporary Latvian Poetry*, edited by Inara Cedrins. Iowa Translations series. Copyright © 1984 by The University of Iowa. Printed by kind permission of University of Iowa Press and Inara Cedrins.

GUNARS SALIŅŠ: Translation "Song," translated by Laris Saliņš, as in *Contemporary Latvian Poetry*, edited by Inara Cedrins. Iowa Translations series. Copyright © 1984 by The University of Iowa. Printed by kind permission of University of Iowa Press, Laris Saliņš, and Inara Cedrins.

OJĀRS VĀCIETIS: Translations "Burning Leaves," "Sunday," "Horse Dream," translated by Inara Cedrins, as in *Contemporary Latvian Poetry*, edited by Inara Cedrins. Iowa Translations series. Copyright © 1984 by The University of Iowa. Printed by kind permission of University of Iowa Press and Inara Cedrins.

IMANTS ZIEDONIS: Translation "How the Candle Burns," translated by Inara Cedrins, as in *Contemporary Latvian Poetry*, edited by Inara Cedrins. Iowa Translations series. Copyright © by The University of Iowa. Printed by kind permission of University of Iowa Press and Inara Cedrins.

LITHUANIAN

HENRIKAS RADAUSKAS: Translations "Star, Sun, Moon," "Hot Day," "Harbor," "Salesgirl," translated by Jonas Zdanys, published in *Chimeras in the Tower: Selected Poems of Henrikas Radauskas*, translated from Lithuanian by Jonas Zdanys, published by Wesleyan University Press. Copyright © Henrikas Radauskas. Translation copyright © 1986 by Jonas Zdanys. Printed by kind permission of Wesleyan University Press and Jonas Zdanys.

POLISH

ZBIGNIEW HERBERT: Translations "Mr. Cogito Considers the Difference between the Human Voice and the Voice of Nature," "Sense of Identity," translated by John and Bogdana Carpenter, as published in Zbigniew Herbert, *Mr. Cogito*, translated by John and Bogdana Carpenter, published by Oxford University Press, and in the United States by The Ecco Press. Translations copyright © 1993 by John and Bogdana Carpenter. Used by kind permission of John and Bogdana Carpenter, Oxford University Press, and The Ecco Press.

GERMAN

CZECH AND SLOVAK

HUNGARIAN

xl

ROMANIAN:

YUGOSLAV

BULGARIAN

YIDDISH

xlii

INTRODUCTION

Eastern Europe is one-half of Europe. It is a vast and vital realm of society and culture which, for all the attention lavished on it since World War II, remains for too many Americans either that sinister expanse "behind the Iron Curtain" or, at best, "the old country," where tales of Dracula, ethnic grandmothers, and good recipes come from. That the region, which embraces ten countries and at least fifteen languages, should be every bit as human and civilized as ours and should also be a source for much of the best in contemporary poetry, may come as a pleasant surprise even to those who are better informed on life and letters in "the Soviet bloc" than is the average citizen. The present anthology aims to contribute to this state of informedness, by way of a fresh glimpse at that genre in East European literatures of which their own reading public has reason to be the proudest—the poetry.

All of us who have come to participate in the volume hope that we are providing a meaningful selection of important recent work—poetry mostly by living writers, "contemporaries" as distinguished from "modern classics"—done in no less than eleven countries, or, fifteen language areas. The order in which these poetries are arranged on the pages that follow is geographic, moving from north to south, with one supranational poetry rounding out the picture. We include verse translated from the three Baltic languages, from Polish, German, Czech, and Slovak, Hungarian, Romanian, four languages of Yugoslavia (Croatian, Macedonian, Serbian, Slovenian), Bulgarian, and Yiddish. We believe in the timeliness of this presentation. The fact that the journal *TriQuarterly,* published at Northwestern University, once had an East Europe number (Number 9, Spring 1967), largely reprinted in book form the following year (*New Writing of East Europe,* ed. George Gömöri and Charles Newman [Chicago: Quadrangle Books, 1968]), does not change the fact that an ecumenical offering such as ours has not been seen on this side of the Atlantic for almost a decade and a half. We think it is time for a fresh beginning.

A sense of a fresh beginning is conveyed, we hope, by our choices; some of the poets in almost every group are not yet forty years old. Our limitations in chronology are simple and straightforward at least in principle: to include the work of poets who were active and influential during the sixties and seventies, regardless of the age of individual poets. Simplicity and straightforwardness were, however, not equally eager to oblige us in practice. When we first conceived the idea for the anthology we argued that much of the older work done in our century is already translated. Duplication was to be avoided, discreteness and at the same time contiguity with volumes printing less recent poetry striven for. This may have been a laudable aim at a time when publishers both here and in

England vied with one another in making available slim selections of the work of individual poets—Herbert, Popa, Sorescu—about whom we were just beginning to learn. I recall with pleasure how charged the air was with excitement, and how everyone felt in possession of instant knowledge about these writers just around our cultural bend. Now, I believe, we have managed to be a bit more sober and practical-minded. Publishing, for one thing, has in the meantime sustained some serious losses, not last the discontinuation of the generous and distinguished Penguin Modern European Poets series. It is also simply a fact of life that many important books in our field go out of print (e.g., *East German Poetry*, ed. Michael Hamburger [New York: Dutton, 1973]). This leaves me wondering whether we can ever seriously insist on showing no more than the tip of the iceberg.

Poets and poetry are, after all, like whole icebergs, continuities; a vital poetry benefits from being displayed within its idiom. In addition, demands on the reader's time being what they are, we simply cannot ask anyone to take in our sampling while remembering or referring to the contents of neighboring collections. A certain amount of the whole has to be available between the same set of covers. Rather than to string yet another discrete bead onto the East European poetry necklace, then, we finally decided on allowing for a certain breadth of helpful overlap. Editorial policy was now revised to read: ours will be limitation not to poets who *came into their own* during the two decades named, but rather to them and to those who worked *and remained influential* during that time. This allowed us to expand our sense of the contemporary and to include some of those "deans among living poets" who went on lending excitement to the poetry scene to the last, some until as recently as little over a year ago. Among East German poets, for example, we chose not to include the celebrated Johannes R. Becher (1891-1958), but we did include Peter Huchel (1903-1981), who suddenly passed away in April of 1981 and whose most recent work dates from 1979. Needless to say, not all such decisions were easy. Vitězslav Nezval (1900-1958), widely regarded as the founder of modernism in Czech poetry, we leave out with the greatest reluctance, while the matriarch of contemporary Estonian letters, Marie Under (1883-1980), who published her last volume of verse in 1963, remains very much with us.

A word should be said about the Yiddish contribution to our volume, a special case, chronologically speaking. Even the most casual perusal of dates reveals that some of these poets are members of an older generation. And yet a closer look shows a full two-thirds of the group, ten of the fifteen names, to be of our time; only five poets—Halpern, Kulbak, Leib, Segal, Ulinover—require explanation of their presence. Miriam Ulinover died at Auschwitz; Kulbak, in a Soviet concentration camp. Halpern, while he died a natural death in New York in the thirties, is regarded as a giant (some of his work has been translated by no less a poet than John Hollander); to

xliv

leave him out and still be claiming that we were representing Yiddish poetry would not make much sense. The same goes for Mani Leib. Jacob Isaac Segal's inclusion seems instructive as one of those critical points where all preconceptions on inclusion versus exclusion break down—his "Candle" and "Rest" are simply two of the finest poems in the entire volume. Perhaps a bit of editorial arbitrariness does not hurt; as futile as it is to try to please everyone, so important is it for editors to show a degree of passionate involvement in what they think they are bringing together.

While age-old distrusts and disaffections among the socialist countries of Eastern Europe are by no means a thing of the past (I would venture, for example, that Lithuanians and Bulgarians come far nearer to being kissing cousins than do Hungarians and Slovaks), East European cultures and attitudes do, in our sense at least, form an identifiable continuum. Partly for this reason, here it becomes much more difficult to know where to draw the line. Ukrainians unhappy at having been left out while we included Baltic writers, will do well to reflect on history as well as geography. In general it was not our aim to include poetries of the Soviet Union, simply because that powerful geopolitical presence, embracing over one hundred languages, deserves attention of its own and calls, in any event, on talent not available to us. But within East Europe proper as here defined, there is unity-within-variety second to none. Languages Baltic, Finno-Ugric, Germanic, Romance, and Slavic, as well as sharp religious and cultural differences, compete for attention and help shape a true spectrum of poetries. Even within this realm there were, necessarily, more exclusions than meet the eye. From among the several folk, minority, and supranational poetic cultures—Finno-Ugric folk and Romany, Lachic and Sorbic, Ladino and Yiddish—we could offer hospitality to the last-named only (*pace* the British poet-translator Keith Bosley, who would have been ready to help us with ample Finno-Ugric folk material). I feel that among supranational poetries Yiddish comes closest to the sedentary condition of contemporary poetry as we understand it. Here too I must record my regret at not having been able to consider such writers as Paul Celan or Tristan Tzara, both of whom the Romanians like to claim as their own. The problem of achievement by one poet in more than one language is a fascinating one; it would be good if someone were to think of an anthology of such poets. It certainly seems much overdue.

A word about some of our choices and opportunities besides those imposed by natural limitations: what we did and did not want, what we could and could not get. We did not want, first of all, to make this anthology an arena for cold-war swashbuckling in either direction; our guiding principles were those of poetics, not of politics. We did want excellent lyric poetry: tight and intense, exhilarating and crafted, unsentimental and engaging. We invited and happily received lyric, and not epic or dramatic, poetry. A poem had to function as a poem; it clearly had

xlv

to have been written for the sake of writing a poem. As one piece of correspondence with a prospective contributor puts it, we were interested in avoiding superpatriotic stances, unobjectified psychoanalytical preoccupations, excessive use of anaphora, and verse so "free" that it shows little or no sense of form or style. We also tried to keep at bay the strident political activist's versified, often declamatory, attempts to air public grievances, real or imagined. The point is not, of course, that justified protest should not take place, but rather that poetry is no place for it. Northrop Frye has nicely summarized the principle in *Anatomy of Criticism:* "...whatever it sounds like to call the poet inarticulate or speechless, there is a most important sense in which poems are as silent as statues. Poetry is a *disinterested* use of words: it does not address the reader directly" (Princeton: Princeton University Press, 1957, p. 4). Were our times closer to those of the New Criticism we could be tempted to invent, for violations of this rule, some such convenient term as "The Heresy of Non-Disinterested Discourse." But even without terms, I think the point is clear. Poetry that "speaks up" ceases to be poetry, not only by dint of the above rule, but also because it becomes, thereby, too much something else. A delicate equilibrium is upset. Good poets have ways open to them of commenting on politics and yet writing a poem. Not many can do it; perhaps such ability is a test of the artist. We stand unbent in our view that direct "rhetoric" in verse is a copout.

In considering home versus exile poets we have shown hospitality to both, although the latter are, for good reason, in the minority. In general we have welcomed the work of expatriates only if it compares in quality with that of Radauskas or of Karpowicz. Some of the poets who have come west are, like these two, highly respected in their home countries. But, as always in art, we are talking about a few outstanding names. It is a logical, although no doubt unfortunate, corollary of prolonged residence abroad that a writer loses touch with his subject matter, if not with his language. Only for the best writers can living abroad become a source of stimulation. Both Mann and Brecht, members of the émigré set of their time and culture, wrote some of their best work in America. So does Czesław Miłosz, the first East European poet ever to be honored with the Nobel Prize.

To pass on from what we wanted to what we could and could not get. If you do not see it, either we did not get any, or we did not receive permission to print it. Those grieved at seeing some of their major poets left out, should have sent us translations. About two-thirds of the anthology consists of acceptances on the basis of submissions (the remaining third being supplied by most of our Consultants), and it was gratifying to have the scores of excellent manuscripts that came in tell us that East European poetry is a vital area of creative and scholarly activity, both here and in the several countries overseas from which work arrived (among them, wonderful to tell, both Israel and Jordan). Chance will, of course, make its input felt, and while the final result is, we think, balanced, blank spots remain here and there. I am sorry to see Slovak poetry underrepresented. Equally unfortunately, we have nothing by the Hungarian Károly Bari, a poet of Gypsy extraction, whose work reminds me of some of the later Dylan Thomas—deeply rooted and experienced nature lyricism expressed in luxuriant imagery and tightly-packed language. It is a shame, but the translations of Bari's work that came in were simply found unpublishable. In some important cases the converse was true: some poetry came in in high-quality multiple translations. The infatuations of fashion tend to govern some of this. No fewer than six contributors sent in versions of poems by young East German poet Sarah Kirsch (who is admittedly very good), while we had to secure on our own the handful of texts we have by Zbigniew Herbert. No more than the number of poems we print by any one poet indicates our estimate of a poet's stature, is the number of poets included per country made dependent on any country's geographic size. We felt that good work deserves to be shown, even by some of the less established poets. We ask those unhappy with any aspect of distribution or weighting to consider that in the space available we had to accommodate the work of many more poets than anticipated.

Translations from fifteen languages are included on the following pages and, there is no escaping it—the anthology has to come across as an offering of poetry in English. Happily, we have been assisted by some of the best people in the field; the editorial masthead and the list of translators speak for the quality of the response that our project has enjoyed. Withal this brings up, once again, that ticklish subject of method of translation: of translators working with or without informants, helpers, trots. The problem of team versus solitary translation invites systematic study that, needless to say, goes beyond the scope of any introductory essay. Let me just say here that, while my own preferences lean strongly to the side of solo translation (and while I firmly believe that there are scholar-poets among us who do a successful job of it), I have here decided, out of need as much as out of inclination, to encourage translators of both species. Strange to tell, there are able teams around, and when a poet himself has the opportunity to check his translator's work, the collaboration can be a happy one.

The translation process is, we might say, the picture window framing the great landscape of which it offers a view, and to readers it is this latter that will prove the more exciting of the two. A seemingly strange land appears, in which we all recognize features of our own terrain. It is almost a truism by now that East European countries, at least those north of the Balkans, found their way into the full light of modern literary history by turning toward the West. Poles and Romanians were always intensely French-oriented; Czechs and Hungarians were, willy-nilly, German-oriented during the Empire, and in the period between the two world wars turned toward France, England, and America on their own. The

Hungarian literary biweekly *Nyugat* (West) (1908-1941) is an excellent example for this militant self-commitment to things Occidental, although it is instructive to note that in 1935 another journal under the title *Kelet Népe* (People of the East) was inaugurated, as it were in answer to *Nyugat*. Indeed the flight of writers from the question of eastern origins, during those heady days of secession and modernism, meant that East European writing pretty much followed developments in France and elsewhere in Western Europe. Symbolism around the turn of the century, along with all the other isms of the modernist phenomenon (cubism, futurism, surrealism, and all the rest), was certainly in full swing in Hungary by World War I, and in Czechoslovakia and Poland as well. Whether we are talking about the various avant-garde groups contemporaneous with Skamander in Warsaw, poetism in Prague, or expressionism and dada in Budapest (and in part also in Vienna), we are observing a shared desire on the part of artists of northeastern Europe of the interbellum generation to be identified as thoroughly westernized—European, chic, up to date.

Up-to-dateness, however, covers a multitude of virtues, and it is refreshing to see that writing in the southeastern countries, and in Bulgaria in particular, does not in meek obedience follow the formula: first comes symbolism, then cubo-futurism, then expressionism, then post-expressionism, and so on. Part of the secret of varied patterns of literary development has to do in some of these countries with delving deeply into ethnic pasts whose values are just now being recaptured. At the same time awarenesses of culture, technology, and writing, whether of western or eastern provenience, are also strongly encouraged. Romanian pride in Roman and Dacian origins, consciousness on the part of Bulgarians that it was their ancestors who gave the Russians and other Slavic peoples their religion and alphabet, thus stand inseparable from a curiosity and love for all that is new and foreign. This very human drive, simultaneously to burrow into one's shell and at the same time emerge from it, makes for an ambivalence that the younger generation of poets have inherited. Ottó Orbán's ongoing experiments in a free-verse form whose origins he can trace back to Whitman offer but one example of this fusion of past with future. Perhaps it is yet another demonstration of the truth that in art there is no progress; there is only self-renewal. The poetry that follows can best show the abundance of energy and talent, the joy in good writing, and the enduring tensions between old and new, eastern and western, that characterize East European poetry at present.

I would like here to express my thanks to all who have helped me assemble this anthology—first to the poets and their expert translators, whose efforts must be seen, within the framework of anthologies such as ours, as affectionate teamwork, demonstrating once again the truth that if the poetry is good in one language it will be good in another. I thank our distinguished Editorial Consultants, all of whom were instrumental in bringing to the project excellent translations and translators; special thanks go to those among the Consultants whose energy and dedication took the form of delivering bodily the material for their sections, often of translating on their own all or most of the poems within the given area. In this regard Mrs. Inara Cedrins, and Professors Ivar Ivask, Thomas Amherst Perry, Jascha Kessler, Aleksandar Shurbanov, and Howard Schwartz have put me in their debt in particular. I am deeply obliged to the many translators, Consultants, and authors and editors of handbooks who provided material for the notes on the poets and introductions to the various sections. In composing the bio-bibliographic notes for which I am responsible, as well as all the unsigned section introductions, my debt to the following standard anthologies, literary histories, and reference works will be all too apparent: *Columbia Dictionary of Modern European Literature*, among other data the source of the section introduction to Latvian poetry; *Contemporary Yugoslav Poetry*, ed. Vasa D. Mihailovich (Iowa, 1977); *Dimension*, Special DDR Issue, Fall 1973, a source of information on the poets Bartsch, Cibulka, Czechowski, Jentzsch, and Kunze; *The History of Polish Literature*, by Czesław Miłosz (Macmillan, 1969); *Hungarian Authors: A Bibliographical Handbook*, by Albert Tezla (Harvard, Belknap, 1970); *Magyar Irodalmi Lexikon*, ed. Marcell Benedek (Akadémiai, 1963, 1978); *Modern Romanian Poetry*, trans. and ed. Donald Eulert and Ştefan Avădanei (Junimea, 1973); *New Writing in Czechoslovakia*, ed. George Theiner (Penguin, 1969); *Selected Post-War Lithuanian Poetry*, trans. and ed. Jonas Zdanys (Manyland, 1978); *Three Czech Poets*, trans. Ewald Osers and George Theiner (Penguin, 1971); *Voices within the Ark: The Modern Jewish Poets*, ed. Howard Schwartz and Anthony Rudolf (Avon, 1980); and *Who's Who in the Socialist Countries* (K.G. Saur, 1978).

Warm thanks go to a number of individuals who have stood by, providing help either in spirit or in substance, during the various phases of work. It was the poet Joseph Brodsky who first told me about the desirability and excitement of an anthology of contemporary East European verse; the many early sessions I had with him, hammering out plans, identifying poets, translators, and sources I must count as basic for the subsequent acquisition of my modest degree of awareness of our vast subject. I wish further to thank the following individuals and institutions for kindnesses too numerous to specify: Adam Bromberg, Deming Brown, Grazyna Drabik, Herbert Eagle, Paul Engle, Eva Feiler, Julianna George, Mary W. George, Renata Gorczynski, Michael Hamburger, Ivar Ivask, Astride Ivaska, Peter Jay, Miloslav Jiran, Rika Lesser, Ladislav Matejka, Seymour Mayne, Czesław Miłosz, Dennis Mueller, Ewald Osers, Marysia Ostafin, Peter Petro, Howard Schwartz, Harold B. Segel, Svat Soucek, Robert Stiller, David Welsh, A. Leslie Willson, the editors of *PMLA, Poets & Writers, Inc.*, the Libraries of The University of Michigan and

Princeton University. Grateful thanks go to all the publishers and state copyright agencies for granting us permission to print or reprint translations, as noted in the Acknowledgments. And finally, deepest appreciation goes to my publishers and friends, Carl and Ellendea Proffer of Ardis, without whose enthusiasm and faith this book would not exist, and who are certainly a part of that non-editorial *we* used throughout this Introduction, if anyone is.

Ann Arbor, Michigan   —Emery George
August 1981

1

EDITORIAL CONSULTANTS

ESTONIAN
Ivar Ivask, Editor, *World Literature Today* †

LATVIAN
Inara Cedrins, Chesterton, Indiana
Valters Nollendorfs, The University of Wisconsin, Madison

LITHUANIAN
Tomas Venclova, Washington, D. C.
Jonas Zdanys, Yale University

POLISH
Magnus J. Kryński, Duke University
Robert A. Maguire, Columbia University

EAST GERMAN
Michael Hamburger, Saxmundham, England

CZECHOSLOVAK
Ewald Osers, Reading, England

HUNGARIAN
Emery George, The University of Michigan, Ann Arbor

ROMANIAN
Thomas Amherst Perry, East Texas State University

YUGOSLAV
Vasa D. Mihailovich, University of North Carolina, Chapel Hill
Charles Simic, Strafford, New Hampshire

BULGARIAN
Jascha Kessler, University of California, Los Angeles
Aleksandar Shurbanov, University of Sofia

YIDDISH
Howard Schwartz, University of Missouri, St. Louis

TRANSLATORS

Fleur Adcock
Elliot B. Anderson
Margot Archer
Rivka Augenfeld
Robert Austerlitz
Stefan Avădanei
Andrei Bantaş
Stanisław Barańczak
Magda Bartošová
Georgi Belev
Bruce Berlind
Mihail Bogdan
Keith Bosley
Katherine Bradley
Bogdana Carpenter
John Carpenter
Mariana Carpinisan
Clare Cavanagh
Inara Cedrins
Michaela Celea-Leach
Allen R. Chappel
Victor Contoski
Stavros Deligiorgis
Marguerite Dorian
Andrew Durkin
Dan Duţescu
A. M. Elliott
Donald Eulert
Laya Firestone
Sylva Fischerová
Lori M. Fisher
Roland Flint
Stuart Friebert
Emery George
Renata Gorczynski
Hellar Grabbi
Eric Groch
Dana Hábová
Michael Hamburger
Miroslav Hanák
Barbara Harshav

Benjamin Harshav
Kathryn Hellerstein
Anselm Hollo
Barbara Howes
Michael Impey
Mark Irwin
Astrid Ivask
Ivar Ivask
Rich Ives
Peter Jay
Peter Kastmiler
Baiba Kaugara
Jascha Kessler
Nicholas Kolumban
Leonard Kress
Magnus J. Kryński
Wayne Kvam
Ilse Lehiste
Seymour Levitan
Irina Livezeanu
Henry Lyman
Edward Mackinnon
Don Mager
Robert A. Maguire
Seymour Mayne
David McDuff
Matthew Mead
Ruth Mead
Vasa D. Mihailovich
Dragan Milivojević
Czesław Miłosz
Virgil Nemoianu
Valters Nollendorfs
Vera Orac
Ewald Osers
Irina Grigorescu Pană
Thomas Amherst Perry
Krinka Vidakovic Petrov
Svetoslav Piperov
Petru Popescu
George Preda

Gabriel Preil
David G. Roskies
Tomaž Šalamun
Laris Saliņš
Lisa Sapinkopf
Boria Sax
Michael Scammell
Laura Schiff
Hillel Schwartz
Howard Schwartz
David Scrase
Aleksandar Shurbanov
Charles Simic
Daniel Simko
M. G. Slavenas
Atanas Slavov
Tadeusz Slawek
Adam J. Sorkin
Mark Strand
Olafs Stumbrs
Irene Pogoželskytė
  Suboczewski
Brian Swann
Timea K. Szell
Tanaquil Taubes
Veno Taufer
R. Loring Taylor
Belin Tonchev
C. W. Truesdale
Elliott B. Urdang
Juliette Victor-Rood
Bronislava Volek
Brenda Walker
Ruth Whitman
A. Leslie Willson
Seth L. Wolitz
Ewa Zak
Jonas Zdanys
Linda Zisquit

East European Poetry: An Anthology

ESTONIAN

INTRODUCTION TO ESTONIAN POETRY

By Ivar Ivask

It seems almost an Estonian fate that events have a tendency to come to late fruition in those northern latitudes. This holds true for both political and literary history. When contemporary Estonian poets search for their roots and beginnings, they can either dip into many centuries of abundant folklore or they can turn to the first serious poet known by name, who appeared comet-like at the beginning of the nineteenth century, Kristjan Jaak Peterson (1801-1822). However, it is equally true that if important events may have a slow start in Estonia, there is no stopping them once they pick up speed. Thus the movement into the twentieth century was spearheaded by the mentally disturbed visionary symbolism of Juhan Liiv (1864-1913) and the prefiguration of expressionism in the free-verse experiments of Jaan Oks (1884-1918). Their hopes for an original Estonian tradition, perhaps even paralleled by an independent state, were more boldly formulated by the poets around the first real school of innovative writers, appropriately called Noor Eesti or Young Estonia (1905).

The Young Estonians asked for sophisticated Europeanization without any loss of what was fundamentally Estonian, because they knew that superficial growth without real roots would not last very long. Neoromantic and symbolist poets such as Gustav Suits (1883-1956), Villem Ridala (1885-1942), and Ernst Enno (1875-1934) gave a new pliability to the verse language and to poetic form. These advances in poetry were accompanied by similar strides in criticism (Suits, F. Tuglas) and language reform (J. Aavik).

It was not until the summer of 1917 that another major school emerged, one which took its name from the mythological bird Siuru. Some of their color and energy can be traced to German expressionism and Russian futurism, but not exclusively so. Just as Suits had dominated the previous school, so now did Marie Under preside over Siuru. Indeed, after the establishment of Estonian independence, which freed the country from the czarist empire and ended centuries of German feudal subjugation, Under became the classical embodiment of the vitality and aspirations of the young Estonian state (1918-1940). She was also a frequently presented candidate for the Nobel Prize, until her death as an exile in Sweden. Critics have compared Suits with Yeats, while affinities were found between

Under and Rilke (whom she translated). Under's work is symbolically represented here with three poems.

In the thirties a third generation matured, the first poets who had received their entire education in Estonian. The distinguished critic and translator Ants Oras (1900-82) presented them in his anthology *Arbujad* (Magicians of the Word) in 1938. Their sensibility was unapologetically Estonian, highly individual, and experimental in differing ways. Three of these poets figure here: Betti Alver, Uku Masing, and Bernard Kangro. Alver has naturally assumed the mantle of national bard after Under's death. Her command of the resources of the Estonian language and her marvelous transparency have been justly compared to corresponding qualities in Racine and Pushkin (A. Aspel). Masing is a mystic prophet, at times akin to Blake and Hopkins, yet with occasional flights into surrealism. Kangro delves into the flora with a botanist's precision and into the national subconscious with the refined tools of a Jungian folklorist. The worship of art and form by Rannit, a poet not included in the *Arbujad* anthology, achieved its greatest refinement during the postwar years. There is also the painter-poet Arno Vihalemm (b. 1911), whose epigrammatical whimsy could have easily figured in this selection had we but had more space. The precedent set by these older poets, for the most part with established national reputations, encouraged the younger émigré writers to make their debut as early as 1946. This is the year when the ironic realist Kalju Lepik, the dialect poet Raimond Kolk (b. 1924), and the startling surrealist Ilmar Laaban (b. 1921) published their first collections in Sweden. Lepik's productivity and many-sidedness was one of the reasons for his selection in the present anthology. The intellectually demanding and erotically frank poetry of the physician Ivar Grünthal represents the émigré poets who came to the fore in the fifties, just as the editor of this selection stands for the poets who made their—often belated—debut during the decade following.

One of the leading figures from the Siuru school who chose to stay home and to collaborate was Johannes Semper (1892-1970). As for the Magicians of the Word group, most of them remained in Estonia and managed to keep their moral and poetic integrity intact by publishing little verse, if any, until the sixties. Alver's first husband Heiti Talvik (1904-1947) died in Siberia and was only posthumously rehabilitated. Alver and August Sang (1914-1969) made a living by translating various classical authors. Theologian Masing's chances of publishing his mystical-hermetic poetry were nil even after a more liberal climate had begun in the sixties, and so he boldly opted for bringing out his poetry entirely with émigré publishing houses in the West.

The example of these "magicians of the word" was as decisive for the emergence of a young, vital group of poets in the sixties as was the independent experimentation of such poets of the middle generation as

4

Jaan Kross (b. 1920) and Artur Alliksaar (1923-1966), who also founded the native theater of the absurd. The presence of these various poetical temperaments helps explain why critics have spoken of a "poetic renaissance" when characterizing especially the achievements of Paul-Eerik Rummo and Jaan Kaplinski. Suddenly there was not only serious competition for the exile poets, but there were also critics who felt that the center of literary creativity had shifted back to Estonia. The artistry and vision of Rummo and Kaplinski can be sampled in our selection.

They certainly are not the only interesting poets writing in Estonia today. There is the virtuoso classicism of Ain Kaalep (b. 1926), the prophetic nationalist stance of Hanndo Runnel (b. 1938), and the ironic clowning, with black-humor overtones, of the even younger Juhan Viiding. Our selection can only be a sampler and a beginning.

In closing, a word about the translations for which I am responsible. I have aimed at fidelity to feeling and imagery, thought, and, to a certain degree, tone, but not to meter and end rhymes. The absence of verbal music is particularly grave in Under, Alver, and Grünthal. Still, Alver's marvelous precision, Under's and Grünthal's profuse imagery are such that they seemed worth recording in themselves.

MARIE UNDER (1883-1980)

Marie Under is considered to be the most important Estonian poet. Her manifold oeuvre (some 400 poems) appeared in thirteen collections, from the first, triumphant, *Sonetid* (Sonnets, 1917) to the émigrée's *Ääremail* (In Borderlands, 1963), spanning a creative life of some six decades. The almost Goethean amplitude of her genius is best indicated through some of her own characteristic book titles, such as *Rõõm ühest ilusast päevast* (Delight in a Lovely Day, 1928) and *Hääl varjust* (Voice from the Shadow, 1927), composed concurrently, or *Ja liha sai sõnaks* (And the Flesh Became Word), her selected poems of 1936. Although attempts have been made to translate her vitally pulsating poetry into the major Western languages, she has yet to be truly discovered as one of the half-dozen leading women poets of our century.

ALONE WITH THE SEA

The rye-sheaves are stacked.
Everybody is leaving.
The carriage-top is up.

The traveller in the back
like the coachman up front
is pensive, silent.

Nobody lingers on the shore
any longer,
not a soul.
It's better that way.
Just rocks and water;
the only tracks are made by my shoes.

The seagull calls.
It's hard. I know why.
The wind rips the water.
And the bee bears from the last flower,
which sways in a crevice,
the final honey.

Thus I walk far
along the white shore
until all of a sudden I see
at my single seeker's feet
that infinite sea. I freeze like a rock.
As if face to face with God, I stop.

(Ivar Ivask)

MORNING JOY

Dawn with a young, restive hand erased
the secret writ of stars from my window,
slammed the doors of night, made more spacious
the dome of blue above the hill of clouds.

The radiant sky kneels in my room.
Sight returns to my eyes, voice to my throat.
The hour beckons to me, a ripe sundrenched apple:
a new harvest has begun from the Tree of Life.

The morning aches to become day in me.
O great newcomer, you! In nocturnal shadows
I became your prophet. My mind's impatient.

Awareness lights the brain. Thought turns keen.
Crystalline light, overwhelm me!
I bear your fiery marks, tearful with joy.

(Ivar Ivask)

AND A STAR FELL

How brazenly new is that old moon up there—
yet so much won't return, for which I care!

O my lost city, o city I lost—
that I seek, always sought at such a cost.

He who has been to the impossible bound,
where is his place? Where can it be found?

Is all myth and illusion? Surface glitter?
A bird awoke in my throat with a twitter.

And a star fell. My heart came ablaze:
what will earth leave me? What will it raze?

Do I grasp nothingness? Is all but a lie?
No, still the main task is left—to die.

(Ivar Ivask)

BETTI ALVER (b. 1906)

The foremost living Estonian poet, Betti Alver, lives in Tartu. After her classical debut in 1936 with *Tolm ja tuli* (Dust and Fire), she devoted herself until 1966 to translation (see, for example, her splendid version of Pushkin's *Eugene Onegin,* 1964). Such later titles as *Tähetund* (Stellar Hour, 1966) and *Eluhelbed* (Life Flakes, 1971) are anthologies which include old poems along with new ones to stress an unbroken continuity of purpose. *Lendav linn* (Flying City, 1979) represents her collected poems from 1931 to 1977. She has also published novels.

FRAGMENT

Great events grow in the shadow.
What doesn't grow in shadow, has no worth.
The victor's power and road-side berries
share the same taste of dust.

Important ideas will rob your peace,
confused ones curl up in pleasure;
the most important ones will never
find form in words.

(Ivar Ivask)

CHILDREN OF THE WIND

A black shadow stalks behind the window; last year it stalked right across
the ceiling.
It always loiters here. But our room is filled by the children of the wind.
Today they roll a cloud-ball into my bed from the shaded avenue.
They are children and simply can't stand the rain talking Latin.
Let it talk! Its words are ruthless. Already I grasp some of them as it
keeps knocking on the roof and speaks about goliaths.
It always invites us to go somewhere—but the door of the hallway is locked.
My food is on a chair by the bed. I'll probably never get back on my feet.

But you fly to places which can't be seen from this window!
Fly gently, children of the wind, over Laiuse Hill.*
Fly over Riisma and Rupsi, over seas of rye and over woods.
From Lake Peipsi fly through Võnnu, where the churchbell spoke.
Fly to Võru, Vändra and Karksi. As you fly across the waters,
look how the Blue Sail looms. Circle around Piibe Road.
Vault over the district of Albu, over Kaarli, Muuga and Ao,
over Assamalla of many ages, over Kääpa and Künnivagu!

Fly across all continents! Leave behind the legend of passivity!
Fly also past success, on your heels Nihil and his twin brother Nemo.
Fly over the sea, on your heels the furies, doubt and the plague,
until you find a cliff. On the cliff is a stone table.
On the stone table is a stone book. The name of the stone book
shines afar from its convex cover:

8

DE GIGANTIBUS

Tarantula and Dragon, Salamander and Skorpion, Homunculus, Horror
and Terror
always consider themselves heirs to the book. Their battle name is Legion.
Always sorrows try to decipher the book. Always the book is razed by
waves of blood.
Always nations hew into it the names and stories of their giant-heroes.
Study that book, my friends, when I can't.
They say that one chapter about giants is entitled:

ESTONIA

(Ivar Ivask)

*For an attempt at decoding the place names enumerated in this stanza, see the
translator's article "Reflections on Estonia's Fate in the Poetry of Betti Alver and Jaan
Kaplinski", *Journal of Baltic Studies* 10, 4 (Winter 1979): 352-60.

STELLAR HOUR

The errant storm does not ask many questions
at life's crossroad.
It's ultimately you who has to answer
for yourself.

However long and dark the night may be,
your forehead still will bear your name.

Even a leaf follows the light, then falls
with others. Still alone.

You lack a splendid goal? Then simply go
and learn how to consume!

Don't you know what slowly makes one gentle?
Why cruelty never comes by chance?
Why flower-helmets do not rust?
Why life's stellar hour comes but once?
Why flickering light persists,
won't die for men on stormy nights?

9

Go ask those better than yourself.
Go ask the living. Ask the dead.

But never ask the passing time
for those who went by chance
into pitchdarkness over Lethe meadows.

Believe me, they don't care
if accidentally or not
the boatman rowed them to oblivion.

(Astrid Ivask)

IRON NERVES

The poet
amidst life's hell implores his fate:
"O grant me nerves of iron!"
    With a sly smirk fate complies:
    "It's a deal! Go in peace!
    Iron nerves you have from now on.
    But, lucky one, your poet's gift is gone."

(Ivar Ivask)

UKU MASING (b. 1909)

Uku Masing, theologian, orientalist, and translator, lives in Tartu. After his first collection, *Neemed vihmade lahte* (Promontories into the Gulf of Rains, 1935), he has published three recent books of verse, all of them abroad: *Džunglilaulud* (Jungle Songs, 1965), *Udu Toonela jõelt* (Fog from the River Styx, 1974), and *Piiridele püüdes* (Striving toward the Limits, 1974). Our selections are from the first and third of these later volumes.

ONLY THE MISTS ARE REAL

The wind is a shuttle made of elm-wood,
I am but an airy web of dusk

Which God's tapering fingers of a unicorn's bone
Wove in the warm room of the stars.
The wind is a shuttle, but of what yarn the woof
On the earthen loom is, I do not know;
The radiance of mists, perhaps, when their power died,
Since my head did not reach to the clouds.

(Ivar Ivask)

FOR POLLA

Oh those big floppy ears which hated flies
and were always full of sand and thistle burrs!
Do they now hear how the dead wolf barks
behind the woods whom your heart always sought?

You can sleep in the sun and not get hot,
thunderclouds won't frighten you into a corner.
Now you run among stars and smell the fog,
lift up your head and howl; the Unknown passed.

No longer you drag yourself through slushy fields,
blind, deaf, with paws that hurt from frost;
you can smell again, you can run along well
wherever you want to, untired by distances!
All fields are verdant now,
no longer must you go astray
helplessly barking in the darkening night.
Now you don't run into stones
and barbed-wire fences
helplessly whining in the infinite night.

You rest on straw now. But in what gardens
does your heart still stroll?
Does it recall that pal of yours
who fell asleep in a snowdrift
with only a snowstorm over his head?
How much I would like to see you again!
Whether you recall me or not;
even if my soul is but a lazy black bird for you
whose shadow bothers you on your ways.

After all, what am I
if I cannot touch your floppy ears?
Oh how I wish you would laugh again
the way I taught you on those winter nights!
You will race up to me at my eternal gate,
not knowing what to do from joy—
to leap right over my head
or slip through under my feet.

I won't have laughed a happier laugh
than on that evening when at last
I'll be where you await me.
I will forget that there was not
a handful of straw beneath me!
I know, and my faith can't be shaken,
in the joy of seeing you again
my heart will grow strong and fiery,
my body will hear my voice as I call it
back from earth and grass,
from clouds as well as the sea.

(Ivar Ivask)

TO EVERY BELIEVER

God's words are golden grains
which a child spilled
when its father took it to the wood
to live alone and to hunger
for the grassy yard back home,
and the rustle of stars
which were at their zenith Christmas night—
singing, the rustle of great waters their melody.

And a white weasel lived under the door.

They roamed the thickets until moonrise,
their soles full of splinters
and their hands full of thorns,
bumps on their foreheads, the body covered with wounds,
but the heart was green laughter.
No matter how much toil and bitter water

they had to drink from muddy ditches,
there was always the taste of a berry
on their lips, sweet and red like sleep.

Until moonrise,
until the hour a yellow ray danced across our path
and the brittle shadows of large leaves
sought from bushes support and protection
against the night's blue power...
And the wind, oh the wind!

And the moon was like the white weasel.

Who can bind, who wants to count
all the scars and welts
with soft hands, the poisons
of the angry and mighty, hidden in braided bread
well before the bloodshot evening!
No one has time at all,
nobody is willing to care or check
whether our veins are still whole.
Because, oh believers, the moon has risen!
Drop your eyes to the ground
like maple-leaves on an autumn morning.

The Lord's words are those golden grains
which no raven cares to touch,
they are like ball-lightning
which frightens the body-snatching demons.

He who goes shades his eyes with his hand
and wonders when he dropped that particular grain.
Should he leave it for others
or take it along in his hair?

Strange, at first, is the road home,
strange the grass and trees,
the spiders' webs which fly over the meadow
right into the water.
With faltering steps, not wishing to go,
one's will torn like the temple curtain,
but then that distant oaktree looks familiar
and behind it the outline of another!
You hurry. Your restive eyes

13

have no fear of extinction,
for who would stay on the road overnight
where demons stalk and frosts watch
each passing flower
to measure its grave?

The white weasel had a tiny cave
right under the threshold.

From word to word silently
the seeker's bright eye moves.
Oh if he only were not overtaken
by midnight's stormy weather!

If he had only placed more golden grains,
his soul could now be on a golden path!
Too often he has to look for them
while his blood trickles gently away
like sand into the fog. The night knows it.
They are nothing but will-o'-the wisps!
Every grain is a harbor, a thousand have to be reached,
past beacons no one has counted—
when we went to the wood.

But the white weasel
watched from the threshold
until the passerby disappeared.

(Ivar Ivask)

BERNARD KANGRO (b. 1910)

Bernard Kangro lives in Lund, where for many years he has directed the
Estonian Writers' Cooperative (1951—), the leading publishing house of
émigré Estonian writing. He has also edited *Tulimuld* (Scorched Earth,
1950—), together with *Mana* the best Estonian literary journal abroad.
Kangro is as prolific a novelist as he is a poet (fifteen titles each).

OLD AGE

One day you will collapse at your desk,
hard at work on a tiring project,
an autumn aster after a frosty night.
Your head so weary, swimming.
Your house engulfed in snowdrifts.
Old age has slipped into the sheen
of copper candlestick and flame.

No hope is left for your escape.
You lift your head from folded arms
and hear a strange and brittle sound.
You strain to listen.

What is it? Someone walking?
Ice splinters into smithereens
under his heavy tread.
Some sort of answer penetrates your gaze.

A magic rod may well have touched
your feverish eyelids, since you see
how all has been but fake,
how all is marked by transience.
You stare into this emptiness
until you startle: Could that be me
there at the desk, humped over
folded arms? that man who's made
of rags and earth?

(Ivar Ivask)

THE TREES WALK FARTHER AWAY

Aspen-blue mist sneaks across fields
right up to me.
The trees walk farther away.

A bird in flight croaks in the fog.
I am brushed by its wings.
The trees walk farther away.

Then the sky turns upside down
and earth ascends to heaven.
The trees walk farther away.

(Ivar Ivask)

LATE FLOWERS, WIND, SEA,
SAND AND FISH

Wind wilts
late flowers,
tiny blossoms
at edge of bay.

Don't blame the breeze!
The sea's there
thundering
upon the sand.

Wave above,
sand below.
Fish laugh
and skip away.

(Ivar Ivask)

ALEKSIS RANNIT (b. 1914)

Aleksis Rannit, art critic and curator emeritus of the Slavic and East
European Collections at Yale University, is the most widely translated
contemporary Estonian poet into English (*Donum Estonicum*, 1976;
Cantus Firmus, 1978; *Signum et Verbum*, 1981). He is the author of eight
books of verse (1937-82).

SO I SEE YOU STILL

Near. Strange. Striding the Rue Royale.
So I see you still. Ever. Towering.

Smiles by da Vinci, inquisitive, curl
in the flickering turns of your mouth.

And for what words? Of you
and all your toil, mere syllables—
veils—but passionate esteem
for Meryon, or Robert Nanteuil.

As though grazing lettered speech
uttered at the verges of your lips,
your voice turned in and spoke to you—
and its flame was buried in your mist.

And now your gaze has dropped again.
But your springing step is free, reinless—
the bustling mobs about you, at your hand
the stone of the December sun—and Paris.

Behind, the columns of La Madeleine
recede with grave and vivid pace.
Cannot Estonian mystique sustain
the cinders of Hellenic sacrifice?

You are silent, and in silence have said
that a craftsman must deny and withdraw,
that craft is strictness of measure,
and temper, and compass, and law.

(Henry Lyman)

WIIRALT DRAWING IN CHARTRES

My lifeless eye and your quickening hand,
my buffalo eye and your swordlike hand.

Swordblade blue, near its Noël Chartres,
your mind, enmeshed in gently blue-pale Chartres.

Swordblade blue, our farewell with cutting frost,
our double loneliness with hibernating frost,

then, taking us from the icy cathedral, Time,
Bach's blazing fire, flame-vivifying us—Time.

Silent in fiery death are eye and hand,
my burnt-out eye, your flame-defying hand.

(Emery George)

JAAN OKS

Net-houses, saunas, barns
let rain slip through
into constructions
of silence withdrawn.

Relapsed. But the recent cry
still passes to the soul.
And silence?

In music
   silence
      is the core.

(Henry Lyman)

BORIS VILDE

> "What I love in music is
> its prelude to extinction"
>   —*from Vilde's prison diary*

When the tender bullet clearly ringing
struck your dawn, no deathstroke now this
only music ringing there. But music ringing
tenderly as thought. A death unclouded, this
your tender bullet clearly ringing.

In name of thought, of one last love the name,
the shot that burgeoned in the air became a rose.
Barbarian fury cut you down, and yet your name

is ringing like the flower called the rose,
in name of thought, in name of love, your name.

Your thought was like the lucid morning
in northern fields by summer seas,
and in you now your new day's morning
is night in dreams of summer seas.
Your thought is of the lucid morning,

the tender bullet clearly ringing.

(Henry Lyman)

KALJU LEPIK (b. 1920)

Kalju Lepik lives in Stockholm. His *Kogutud luuletused* (Collected Poems, 1980) gathers poems (1938-79) published in ten collections, all of them in emigration. *Death Has a Child's Eyes* (1976) contains some of his poems translated into nine languages. He has edited the *Collected Poems* of symbolist poet Gustav Suits (1963). His contemporary adaptations of folk song techniques have influenced Estonian poets at home and abroad.

THE SEA

1
When the sea can't bear me,
I don't walk on it.
Give me something, sea,
to put on my table.

We ate a dangling sea-tail.
It sure was salty.

2
The sea under my fork on my table.
Under the fork's prongs
the head of the sea and seven eyes.

The seventh eye
foretells stormy weather.

19

3

You sure have salty eyes, sea,
and a blue nose.
White is your handkerchief
when you blow your nose.

There is a roar in your ears.

O my sea,
you have broad hips
and bouncy thighs.
O my, how you carry on,
my sea.

You lick my toes
with a salty tongue.
The sun strokes
your naked belly,
as I sit
on Roslagen's rocky shore.

But why do you suddenly spit
into my face in October
when I was born?

(Ivar Ivask)

LET ME HOLD YOUR HAND
   For My Daughter

Let me yet hold your hand
though day has gone,
night with its lantern come.

My hands are earth-rough.
My lips wear war-scars.

A father's mouth nevertheless.

(Ivar Ivask)

CURSE

If you destroy our tongue.
If you destroy our people.

May rain turn to stone over your fields.
May stone shoots sprout from stone.
May stone bread be on your table.
May stone be the soil under your feet.
May stone be the sky above you.
May the sea turn into stone.
Into stone as your heart is of stone
against our land and people.

(Ivar Ivask)

NOBODY TALKS EARTH'S LANGUAGE

Nobody talks earth's language.
Nobody knows sand's language.
Nobody chats in the language of clay.

Out of earth you were made.

How will I talk under earth?
How do I chide my wife?
How do I scold my children?
How do I sing under sand?

Earth cannot ever grasp earth.

(Ivar Ivask)

IVAR GRÜNTHAL (b. 1924)

Ivar Grünthal, a practicing physician in Göteborg, founded and edited the avant-garde émigré literary journal *Mana* (1957-66; continued by Hellar Grabbi). A critic and translator, he has authored six collections of verse (1951-64) and two intricate novels in verse. Our selections are taken from his third book *Must pühapäev* (Black Sunday, 1954).

A CONSTANT STARBURST

A constant starburst showers us,
nights's bridges snap now one by one,
and him who begged for Hades' mercy,
the flaming skies turn into stone.

The Prince of Darkness moved a notch,
his wily hand was slowly tamed.
Yet Northern Lights are still a threat
and moonbeams do ensnare us.

We only wished to be alone
in evening's darkened shelter.
But stellar quicksands buried us,
and hosts of angels full of rage.

(Ivar Ivask)

I BROKE THE SECOND PART

I broke the second part of the First Commandment
and wrapped in fir-twigs the entire sun:
a black and thorny tree kept growing
right through the blazing freedom of your flesh.
Why were you silent, when with hands of stone
I slowly shaped you into bird and fire?
In forest-marshes all my gods then perished,
because you burnt them with extended wings.

(Ivar Ivask)

ALL GREAT MEN

All great men were ground to powder
who once made stones cry forth.
They who took wreaths to monuments,
were crossed off black lists themselves.

Death lives from hand to mouth on Cathedral Hill;
two eyesockets stare from sacrificial stone.
Blood of the lamb is replaced by rain;
suicide is strictly forbidden.

The empty seats on Devil's Bridge
laugh openly at Liberator Alexander.
The leafy chestnuts have no interest in pursuing
with moist eyeballs what young lovers do.

(Ivar Ivask)

AND THUS A MONOLOGUE

And thus a monologue keeps flowing
in a thousand lakes of memory—
Karelian blue.

The faces of the dead fell
from the nocturnal sky like stars
into barren April snows.

The black bullet-wound between the eyes
has barely scarred the skin
as moonlight flows from their faces.

Dust does not turn to dust.
Ice will be adrift
once night turns to dawn.

(Ivar Ivask)

IVAR IVASK (1927–1992)

Ivar Ivask is professor of Modern Languages and Literatures at the University of Oklahoma, where since 1967 he has edited *World Literature Today* (formerly *Books Abroad*) and has directed the biennial Neustadt International Prize for Literature (1970-). He is married to the Latvian poet Astrid Ivask. He has published six collections in Estonian (1964-81) plus *Gespiegelte Erde (Mirrored Earth,* 1967), written in German. *Elukogu (Life Collection,* 1978) is an anthology of his poems from a twenty year period, illustrated with his own drawings. His first one-man show was in 1979.

FOLIAGE FOLIAGE

Foliage foliage
autumn wind constellations
clarify nothing for me
nothing changes but shrill voices
punctuate unspoken sentences
unfold on the wind
carry strangely
hourly for the owl
the years unfurl
but it does not count
fish-like the soul
slips into the coolness of fall

(the Author)

AND THEN I TOLD

And then I told about the red soil of Oklahoma
bloody clay with scooped-out artificial lakes
truth visible to the bird's-eye view of a plane
walking in the streets one doesn't see death
that featherless blood brother of the Indian
a plucked rooster on the blood bath's wall
no brave eagle King of the New World he
In Andalusia Crete earth is blood-hued too
yet into the olive trees has risen
all slavery hecatombs forgotten peoples

In the earthquakes of history the Alhambra
persisted somber hedges of myrtle
in glinting water *Patio de los Arrayanes*
And then the proud broad stairs in Phaistos
which now lead to Italian archeologists
whose swarthy native help swallow
Aphrodite Athena Adonis figurines
the earthen art of yesteryear warms in the belly
(we too bought a coin which had emerged
from the intestinal labyrinth
I carry it on me always)
Here olive trees grow like aging men
into whorls the world is still atom-old
death is a faithful wife weaving at home
until weary-wristed Odysseus lands
all continents seen all women tasted
but not tasted death on his own island

(the Author)

WE SHALL RETURN

We shall return   we will stay here
                              we will all die out
there   here
               that should only be a starting point
for a seagull   for man
                    who flashes back and forth
across the infinite sea   from continent to continent
                    caught in the net of memories
          the bullet of a better future through the heart
          the present is always differently present
                              but how
we will stay here   we shall return
               a game of black-and-white marbles
               seas and continents and history
                    speak another language
                    other truths travel by sea
          in my window several ships ride at anchor
where do they come from   where do they go
                    I don't know
          the circling seagull surely knows

25

                    the crossed masts of the sailing ship
black   white
                    against the sea's harshly burning truth
                         the sun's full stop
the beginning of life   the threshold of death
                    a black-and-white game
                what's the use when heads roll
                    rocks on the shore
                history chases them across the waves
here   there
            the bullet of a better future in the neck
we won't stay anywhere   we won't go anywhere
                a solution right here and now
                tension between several poles
                     is that the only way
                caught by memories ideals

                Other truths travel by sea
        in each window several truths ride at anchor
            all leave in every direction at once

(the Author)

I AM THE SHEPHERD

I am the shepherd of the heavenly flock
a white infinity of sheep is below me
what sheared curly hills of wool
I am a humble and quiet spirit
on the snowfield of the endless North
a flaming blue sword
leaves my hand a swan
with the fingers my limbs
abandon me
I descend to the eternal snows
to brood on the world's egg
I lean against the pale shell of the visible
holding on to the North Pole
giving up the ghost
I sway in the space of space
icier than an icefloe
I run back into myself

26

fleeing from the radiant peak
on which vision
fades into invisibility
no one can deny me
to vanish into my own tongue
a seed of seeds

(the Author)

JAAN KAPLINSKI (b. 1941)

Jaan Kaplinski lives in Tartu. An influential essayist with a wide range of
interests (ecology, minorities, linguistics) and a translator, he has published
six books of poetry (1965-82), the most celebrated being *Tolmust ja
värvidest* (From Dust and Colors, 1967), which introduced new modes of
expression into Estonian poetry.

WHITE CLOVER WILL NOT ASK ANYTHING

white clover will not ask anything
but if they should ask in whose name
I answer in the name of white clover

bones and brass buckles were left on the soldiers
resin blotted out the crosses from the pines
white clover white white clover

one bare branch three leaves father son and holy ghost
dark pine needles bark being torn by the wind
red color was the question green color is the answer

(Ilse Lehiste)

I UNDERSTOOD I UNDERSTOOD

I UNDERSTOOD I UNDERSTOOD
"is beautiful" is an illusion. It is time for esthetics
to die. The moon sets. Categories
lose their meaning. Something new

grows through the walls of Sparta.
JERICHO
blow your trumpet John Coltrane blow don't be dead
return be a revenant be a phantom only don't be
silent shriek Ray Charles don't forgive Archie Shepp
blow blow away the firm cities the memorial tablets the holy books
the national heroes classical literature the renaissance the epics
the romanticism the Young Germans the Slavophiles Cromwell Richelieu
James Cook Columbus Vasco da Gama Philip Louis all of them
blow away their discoveries their borders blow
away their names their rooms and streets away away Ludwig van
Beethoven G.W.F. Hegel Goethe Disraeli Alexander by
the grace of God Johann Strauss Baudelaire James Joyce away into the
wind
of oblivion into the hot holy black wind of oblivion
their philosophy their music their pride and history
blow their banners inside out their moneybags genealogies
memoirs museums monuments of art tapestries draperies
capitals parliaments parties away their culture their
armless marble statues away back into the earth pantheons Phidias
Praxiteles broken pale statues which for centuries have
profaned the free living earth reviled the children and the sparrows
from their places up high blow into the burning all what is hope
black coal in the hard palms of rock layers seeds
in the black womb of the earth the color of hope is black this is hope
what survived history Augustus Columbus Vasco da Gama
the passage from the Slave Coast to the shores of new worlds hope is
what helped the living flesh to believe to conquer chains and Calvinism
hope is a song the rhythm the knock of waves breaking the testament of the
drowned
the screams of ashes the revenge of the Indians the living flame the
drummer's
fist always without the past always forever living SANCTUS SANCTUS
SANCTUS.

(Hellar Grabbi)

LIKE A BLACK HEDGEHOG

Like a black hedgehog
eternity descends
into the valley

28

in the lap of a child
a barbed ball
melting open
alive

the world's boundaries
the world's barbed wire
moving
like hedgehogs
across boundaries

children's eyes
like butterflies
resting
before
you
on the ground.

(Ivar Ivask)

IN SPRING DURING SPRING

In spring during spring
the difficulties were easy dearest
does that which we bore bear us
faith hope and love
like waves Monday Tuesday
Wednesday Thursday Friday Saturday Sunday
return paying back to our head
unnoticed hours unslept nights
ebb-tide corals living fish and rainbows
of the South Sea of Ihaste Urvaste
Lawrence we too burn under us too is a fire
Monday Tuesday the rainbow comes to an end
at one end a bull's head at the other a writer's bones
a poet's bones with home wife and kids
a coffeemaker which was not to be forgot
which you forgot to unplug a burned coffeemaker Larry
we burn we are coals time ashes
because fire won't win ashes will
mouthless footless eyeless handless ASHES
in our mouth feet eyes hand dearest

still you are a red rooster
still I'm a red rooster
LAWRENCE COCK-A-DOODLE-DO.

(Ivar Ivask)

PAUL-EERIK RUMMO (b. 1942)

Paul-Eerik Rummo lives in Tallinn and is married to the poet Viiu Härm. Author of several seminal plays in the absurdist vein (*Cinderellagame* was produced in 1971 at New York's La Mama Experimental Theatre Club), he has also translated Dylan Thomas, contemporary Finnish poets, and others. His verse from 1960 to 1967 is collected in *Luulet* (Poetry, 1968), which contains his most famous volume *Lumevalgus... lumepimedus* (Snowbrightness—Snowblindness, 1966). A more recent book remains unprinted because the poet insists upon its publication in its original form.

HERE YOU GREW UP

Here you grew up. On a land which is flat.
You get your peace and balance from that.

The Egg Hill remains the cloud-frontier.
The clouds are low and mouse-grey here.

One ticket to the world was meant for you.
You still can check whether it's all true.

What Mecca is to a Moslem believer,
these woods are to you with mushroom-fever.

Here you were born. On a land which is flat.
Your peace and balance stem from that.

(Ivar Ivask)

AGAIN AGAIN AGAIN

Again again again again again
As soon as I close my eyes
it happens once more
that which a beekeeper told me
about twenty-five
beehives
that burned down
This truly happened story
This truly dead apiary

This is not a poem. The subject matter must crystallize in order to become a
poem.
I have waited, but it does not crystallize and does not crystallize, and I can
wait no more. This is not a poem. You seek from a poem meaning between
the lines, hidden thoughts, symbols. There are none. This is not a poem—

This is a rake with fiery tines
which rakes the heather to cinders
this is a saw with fiery teeth
this is the gnawing through of the beehive's feet
this is the melting of the honeycombs
this is this that honey does not extinguish fire
this is this that even honey catches fire
this is this that the bees sting the blazing air
sting the air
and that they perish
This is the swaying of ash-butterflies

Then from far away returns a scout
to dance for the others the news
that somewhere he found blossoms never seen before

This is this that I do not know
how to make clear to him by dancing
all that happened here
and how to answer him
when he asks
why all this?

(Ivar Ivask)

31

THE SKY BENDS

The sky bends over the earth
like someone who picks potatoes.
The earth clings to our feet.
It really appears to love us.

(Ivar Ivask)

THE WORLD DID NOT FORCE ITSELF

The world did not force itself into my soul, it seeped into it,
it did not break into my heart, but corroded it.
It stayed long waking nights in the corner of my room,
a radiance around its head like a halo.
Did it want to torment or console me,
shortening on purpose the wick of my sleep-lamp?
Rock me? Irritate me? Soothe me? Upset me?
Obligate me in any way? Well, did it want
anything at all? Perhaps at an hour
when my pupil contracted infinitesimally,
it entered like a passerby in search of shelter,
rested awhile and, sensing that it was enough,
girded itself and departed without further ado—
just as it had come some time ago.
It left me to breathe its unmawkish kindness,
freshness without bite, an even fire,
a fire that does not destroy, a colorful sobriety;
a voice, everywhere for a brother, a sister;
bright indifference; the all is indeed
one—all one—alone.

(Ivar Ivask)

JUHAN VIIDING (b. 1948)

Juhan Viiding, like Rummo the son of a well-known poet, is a versatile
actor of many masks who lives in Tallinn. Under the pseudonym Jüri Üdi
(George Marrow) he published five collections (1971-75) and then made an
anthology from them under the title *Ma olin Jüri Üdi* (I Was Jüri Üdi,

1978). Strictly speaking, *Elulootus* (Hope for Life, 1980) is the first new book to have appeared under his real name.

SPEECHES IN THE INTERIOR

My leafing of myself is like the casting of one's own fortune
        in tin; like pouring lenten oil into a fire of wood shavings;
it's a trial—against myself, for myself.
All that I do is an endless overture,
the lifelong drawing of a lifesize picture
while jailed for life;
preparations to reach the interior.
There in the interior I will give those speeches,
onetime speeches on freedom.
And only those who will listen to me until the end will be free.
I will give them my freedom
which flew right through me.

(Ivar Ivask)

I CELEBRATED THIS DAY IN MARCH

I celebrated this day in March
In such a perfect room that I lacked comparison
all comparisons are frequently so wrong
and knowing this why compare
from this comes strong imagery
from which no single image can be detached
what's the use of the poem's trying
there's no point in just trying.

(Ivar Ivask)

GEORGE MARROW'S 1011TH DREAM

Let's put the cranes in an insane asylum for observation!
Psychiatrists, learned men, make these birds accountable!
Although they have felt no cold, they are able to think of winter:
command your male nurses to fly up after their formation!

Let them bring all the little birds and animals in for examination,
set snares for mice and rats and other rodents!
Inject them with insulin to put them in a coma
and in the process they'll confess to all their tricks.

How do they know to head for shore when the ship's not sinking
but only leaking and safe dockage must be found?

Foxes, those chicken thieves, should receive chicken therapy,
but the pike should get anti-perch therapy.
Investigate the bunny rabbit for schizophrenya,*
the bees for melancholya, the ass for having caught an epidemic
or something else. Check the striped zebra
for a split personality or perhaps leprosy;

find out whether elephants suffer from megalomanya,
if the doe has chronic fear, the buck is paranoid,
if the squirrel has a phobia for green and pink,
whether the silly kangaroo has not turned into a hypochondriac.

In the zoo the wind blows and it rains cats and dogs
when along comes the storm's main attraction—a flying birdcage.
In Mr. Marrow's nocturnal brain the squirming snake spake:
You crawl, I crawl. I am the redeemer.

(Ivar Ivask)

*) The poet's alteration in spelling of this and some subsequent nouns is intended to give
the words a more Slavic flavor. I.I.

I AM A SERF

I am a serf. I work in the sweat of my brow
for the sake of the landlord.
It sure is bad to work for the landlord.

The work is hard, the overlords are foreigners
and don't understand that I am a human being.
I sing ritual and work songs, especially work songs.
These songs are transmitted orally from one generation
to the next and are a part of folklore.

I often wonder whether the peasant's life in Germany
is any easier than mine. But I don't know.
And whom would I ask? The feudal landlord?
Don't make me laugh.
He doesn't know my language and wouldn't understand
    my question.
And even if he knew my language and I were to tell him
that someday there will be books written in Estonian for sale,
he probably would let them give me countless whiplashes
in the estate stables.
And if I were to add further that eventually there will be
an independent Estonian theater, he would laugh in my face.
But I am not going to tell him that.
I feel the time is not yet ripe.

(Ivar Ivask)

LATVIAN

INTRODUCTION TO LATVIAN POETRY

By Valters Nollendorfs

In the wake of World War I, the emergence of the USSR and an independent Latvia caused a number of splits. A group of Communist writers in exile was formed in the USSR, but its most prominent member, Roberts Eidemanis (1895-1937), and many others perished in Stalin's purges. Sympathizers who stayed in Latvia fared better. Indeed, that was the time when literary activity reached a high level in Latvia; the ratio of published titles to the population was among the highest in the world.

Although expressionistic tendencies appeared briefly after World War I, particularly in the poetry of Sudrabkalns and Pēteris Ērmanis (1893-1969), style consciousness predominated, especially in the classic, elegant verse of Edvards Virza, pseud. of Edvards Lieknis (1883-1940) and, later, in the poetry and prose of Ēriks Ādamsons (1907-1946). Formalism with an ethnic twist is displayed in Jānis Medenis's (1903-1961) attempts at neoclassic Latvian meters. The ethnic-national component remained strong and was encouraged, particularly during the authoritarian Ulmanis regime in the late thirties. "Positivistic" tendencies were, however, opposed by post-expressionists and Marxist social critics, among them Sudrabkalns and Laicēns, but most prominently by Aleksandrs Čaks, whose poetry celebrates the city with its underprivileged people. The formal characteristics of his verse—its rhyme, rhythm, and imagery—have influenced succeeding generations of poets.

World War II and subsequent political changes disrupted many writers' lives and work and opened a chasm between the literature produced in Soviet Latvia and that produced in Western exile. Those leaving Latvia at the war's end include not only established writers, but also many whose activity had just started, such as the lyricists Zinaida Lazda, pseud. of Šreibere (1902-1957), Andrejs Eglītis (b. 1912), Veronika Strēlerte (b. 1912), and Velta Toma (b. 1912). The works of Lazda and Toma are steeped in the folk idiom; Eglītis is best known for his forceful patriotism; and Strelerte's forte is her fine lyric balance. Once abroad poets as well as novelists continued to write in Latvian, thus having to face a dwindling audience. The younger generation of Latvian writers in exile is, for the most part, poetry-minded. In the fifties such poets as Velta Sniķere (b. 1920), Dzintars Sodums (b. 1922), Gunars Saliņš (b. 1924), Linards

Tauns, pseud. of Arnolds Bērzs (1922-1963), and Olafs Stumbrs (b. 1931) started to display literary characteristics steeped in Latvian and foreign traditions. Although the sixties produced such new poets as Baiba Bičole (b. 1931), Aina Kraujiete (b. 1923), and Astride Ivaska (b. 1926), it was primarily a decade of prose. The early seventies witnessed a new shift to the lyric when the generation of writers born in the fifties began to publish.

For more than a decade after World War II, literature in Soviet Latvia adhered to the officially prescribed tenets of socialist realism. Liberalization after the post-Stalin "thaw" resulted in new critical attitudes and some formal experimentation. The lyric genre showed the greatest vitality here as well, particularly in the work of Ojārs Vācietis (b. 1933), Maris Čaklais (b. 1940), Vizma Belševica (b. 1931), Imants Auziņš (b. 1937), Jānis Peters (b. 1939), and Imants Ziedonis (b. 1933), the last-named poet being the most versatile of them all. The basic tendency among the lyric poets is to develop fully the expressive range of the poetic language itself, absorbing, specifically, colloquial and idiomatic elements to create verse that is simultaneously contemporary and steeped in tradition. Despite relaxations, literary activities are still monitored. Although visits by exiled writers to Soviet Latvia are encouraged, exile literature, with few exceptions, is not permitted to circulate freely, and strict separation is the rule.

ANDREJS EGLĪTIS (b. 1912)

Andrejs Eglītis studied at the Technical Institute at Riga, and has lived in Sweden since 1945. He is Secretary-General of the National Latvian Fund, and is also one of the editors of the émigré magazine *Ceļa Zīmes* (Roadmarks), published in Sweden. He has to his credit over one dozen books of poetry and prose, and has lectured in Australia and in the United States. The poems below are from the cycle "Otranto," written on a journey to Australia.

WINTER...

Winter has already shriveled—spring rushes in with
                    black burial steeds.
Draped in the homeland's forests, sunlit mists veil
                    the hilltops—
I wait for floods to bring back the ruins
                    of cathedrals and congregations—

38

the footbridge thrown onto last year's shore no longer
                         reaches across, stops at the depth's center.
On the southern slope, forest doves begin a sober
                         calling,
Calling—will the waters rinse a clear heart's
                         beloved guests now in death's domain?
Those who have lain in foreign snows, long all the more
                         with spring's coming—
Decaying forgotten, long to be uncovered to flowers,
                         sky and the flesh of birds.

(Inara Cedrins)

IN A NAPLES BAR

In a Naples bar, the Virgin's sooty face.
Obscured by the flicker of an oil-dipped wick.
Morning has dawned. A cliff climbs slowly downward
                         from the clouds—
We are still going through a dark soul's night.
What are you doing in a raucous bar, Blessed Mary?
     Think also of us.

(Inara Cedrins)

BLACK DOVES

Black doves with white breasts walk the church in circles,
Nuns light candles, cheeks hidden in veils.
Faces of holy images grimace in somber warning of death.
Candles glimmer in green emerald; as at home, when chestnuts bloom.
Lotus smoke is dizzying, covering worldliness with a veil—
Like a mist to the heavens—with grandfather, riding the nightwatch.

(Inara Cedrins)

I SEE HOW

I see how the grasping fingers of lovers do not let go, till
                          day installs itself, and
                          the world's droning.
The ocean shakes off its slept-away dark dreams onto waveheads
                          in the sparkling pale light,
Morning tears through cerements in a rosy sky—
                          most precious will be
        what fingers, letting go for eternities, give.

(Inara Cedrins)

AINA KRAUJIETE (b. 1923)

Aina Kraujiete, who is poetry editor at *Jaunā Gaita,* has five books of
poetry out in Latvian; these poems are from *No aizpriktas Paradīzes* (From
a Bartered Paradise, 1966). She writes a free verse whose chief
attractiveness lies in its flexibility and emotional and intellectual balance.

CITY GIRLS

The city's pale-blooded
girls stretch up high
in windows and
with fingers delicate as Fijian chrysanthemum petals
toss them
shut and open.

Their figures wouldn't leave gentle impressions in beach
sands, and goose-down softness does not await them
in a twosome joy.

Nervous, as though disturbed from over-sure dream, they
feel of woven flowers in the curtains and listen tensely

how steps reverberate
scuffling
on the stairs—
        stride and—
                slide off

past the door. Passionless
their mouths like pale
thistle blossoms stick softly to the palm,
so as not to become loud.

Unnecessary
to themselves, and useless to the never dark boulevard,
the city's pale-blooded
girls wilt in the windows and
toss them
shut and open.

(Inara Cedrins)

IN CHALK ROOMS

        let all the walls
        about me be green
        like a whole harvest
        of summer's meadows

walls like light
    sliding through green bottle-glass
walls like forest
    moss-grown and damp
walls like mouldy
    cheese on a knife
and walls like frogs
    so cool and loud
walls with gentleness
    like budding leaves
walls of juiciness
    as of chopped turnip tops
walls in that tone
    in which rain soaks moss
and yellowed like cabbage
    butterflies abandoning cocoon
walls hard and green
    corrugated wet
encompassing me
    as have only woods and waters.

but the world listens
to me as to a gnat's song
though I suffer terrible
famine for greenness

(Inara Cedrins)

VAGABOND'S ARISING

—slide and—

                      pass on
past the door
his wavering carefree steps.
Raising up dust in the corridor.
        Come to rest—
to reverberate again in the night like pebbles
on the stairs and at the entrance
to the city, upon which as on an immense sink
ordinary people flow together
as flood waters whirlpooling
buildings, skyscrapers, underground regions,
and then rinse out again to go forward on the streets.

In the hustling crowd like a button torn off a coat
underfoot for everyone, unnecessary, he goes sensing
—tonight he should be far from this place—

but all his senses
yearn for strange smells and sounds, strange pleasures,
leaning to the sensual like grass to the rain,
and not able
to turn away any longer from lit window—
as though by some miracle
seeing in it

the Oberammergau Passion Play
or sinning Franciscan monks at vespers
or perhaps
something else. He doesn't know—
               will the bells ring
of Christ arising or toll his own death;
if only he could

42

go in, where there's light and
                stay—

In the morning, when droplets of rain
spray like dew on
his face (yellow and sour as a lemon)
he awakens—
not even on a bench,
but in place, where in the cracked asphalt of pavement
the damp tar will not allow a single shoot of grass

—where no daffodils will bloom in March—

and the holy morning's rumbled over by everyday steps.

(Inara Cedrins)

GUNARS SALIŅŠ (b. 1924)

Gunars Saliņš is author of several volumes of poetry and a recipient of
the Literature Prize of the Latvian Cultural Foundation of North America. He lives in New Jersey.

A BOMB VICTIM IN THE CATHEDRAL SPEAKS TO A CHILD

Munich, Spring 1945

How will you crawl in to me, little one?
Along halftumbled tower foundations,
playing with shards of colored glass?
Don't cut your hand
on the broken smile of St. Ann!

A wing tip?—From the holy spirit, likely
(it is now somewhere close to us, quite close,
the holy spirit—perhaps if you dig,
you'll find it sometime).

And many, many more such things lie here,—
if you consider, here all things are holy,

only crumbled
and twisted
and fallen in tangled heaps.

Sometimes it seems to me
that even I, here lying, become changed,
and shall some day be able to provide
you with true miracles:

I will appear to you
above the ruins, transparent and mild-colored,
as if painted on invisible glass.
You'll gaze at me and through me then
into the heavens and paradise—
and then will tell me of it all,—

for I myself will be but like a window,
with colors mild, transparent, but blind.

(Baiba Kaugara)

SONG

In the cold my voice grew hoarse,
and one day my song
had frozen solid.

I drank hot milk with honey
and recited a single prayer:
for the song to return—if only like
the mooing of cows or the humming of bees.

Then it happened: one night, when all
who tended me were gone,
from the slaughterhouse on the waterfront
cattle broke into the streets.

Thirsty, they mooed the city full. Galloping,
with their hot breaths and bodies
they melted the snow from windows and trees,
from skyscrapers and squares.

I pushed open my window—a scream
thawed my voice. From it, as from a river,
cows were drinking at noon—dipping
their warm udders in my song—

their warm udders in my song.

(Laris Salinš)

ASTRIDE IVASKA (b. 1926)

Astride Ivaska attended the University of Marburg, where she studied
classical, Romance, and Slavic languages. She is married to the Estonian
poet, and editor of *World Literature Today,* Ivar Ivask. Mrs. Ivask has to
her credit three volumes of poetry in Latvian, and she has completed a
volume of prose sketches. Our selection of her poems below comes from
two of her books, *Ziemas tiesa* (Judgment of Winter, 1968), for which she
won the Zinaida Lazda Prize; and *Solis silos* (A Step into the Forest, 1973),
for which she was awarded the Latvian Culture Foundation Prize for
Literature. All but one of the translations are reprinted from *At the
Fallow's Edge,* bilingual edition with translations by Inara Cedrins,
published at Santa Barbara in 1981.

FOR MY GODMOTHER

At times this light comes to me again
and asks: who are you and why?
And I answer, sometimes in childhood
on my godmother's farm in the evening
after strawberries and cream we watched
shadow plays, which always ended
in one and the same place—
before the Great Wolf put
to Kasper the question: who are you
and why?
More next time, we were told,
but another evening it started from the beginning
and stopped just there,
where the Great Wolf had almost
grabbed Kasper in his teeth, but not quite.

Candle blown out, we drowsed
in the low board room above the stables,
with the scent of hay, window down to the floor
and the moon throwing shadows on the ceiling.

(Inara Cedrins)

From AUTUMN IN THE CASCADE MOUNTAINS

1.
Wild ducks had flown, and early this year
the wind began roaming tree-tops,
dividing living leaves from dead.
At a turn in the winding path the valley opened,
waterfall staining the mountain's face—
suddenly I knew life by heart.
Wild ducks left early this year,
winter will be severe,
you'll never again be as light and clear
as now when I look at you.
Perhaps you'll remember autumn, the bay
where beating waves sing, perhaps
you'll remember it as a long-ago tale,
when you darken, and the valley
turns dusky below your feet.

2.
You come by ancient winding paths
from the mountain's heart, slowly
as a miner climbing to light you come
carrying your tiny swaying lantern.
From ancient and abandoned levels
you come toward me, and in your eyes
flicker briar-roses.

3.
It's long since anything was so light
or cool; so gentle
and still, and yet so close,
so deep and cutting

that there is nowhere to flee,
nowhere to flee.

(Inara Cedrins)

K.H.

You were the first to leave
of our summer friends.
Between your house and the granary
are still woodlands—
slender birches and the junipers'
bristly fur coats
against fresh pine growth,´
tight in the darkness.
Strawberry beds relax under the snow,
the hawk that was hung up as scarecrow
stiffens. Life has lifted off on waxen wings.
The lathe drowses in the granary,
unused blocks of birch wood
feel about with their blind eyes.
On this island frozen in snow, twilight settles.
Each day at this hour
on the sills, cabinets, shelves,
the cranes
stretch out their necks and softly
begin calling in the silence.

(Inara Cedrins)

TO THE MEMORY OF A POET

This spring the cuckoo won't count out your years,
Naomi, nor will the Rain King invite you
to walk over the sea, patting wave backs.
This spring we will leave you
in your green-glass palace,
alone behind many blind windows.
Your radiance will unfurl
around blossoming chokecherry branches,

and you'll no longer fear
tide nor ebb,
nor the moon's changing moods.
We won't tell you about the festivities
for which we were so solemnly preparing,
when we came upon you—trodden bloom,
sprig of lemon, verbena on the garden walk.
As a broken sunflower we found you,
Naomi, and we wrapped you up
in the Rain King's veils,
asking in whispers
what sort of wasteful creator is he,
who left you on the garden walk.
We placed chokecherry flowers
over your light-grey eyes,
and forbade the cuckoo to call,
so that you might dream, Naomi,
of your life's miraculous chance.

(Inara Cedrins)

THAT WHICH HAS REMAINED UNLIVED

to the memory of a poet

That which has remained unlived,
in this world unlived,
will hurt in the next.

Cloudberries ripen in the marshes.

That which has remained unsown,
in this world unsown,
will wither in the next.

I have no basket.

That which has remained unsaid,
in this world unspoken,
will weep in the next.

Cloudberries freeze in ice.

(Inara Cedrins)

VIZMA BELŠEVICA (b. 1931)

A major Soviet Latvian poet, Vizma Belševica often writes of a world in which nature (plant growth) and art (the specifics of Western painting) join man's everyday concerns in the big city. Two of her books, *Jūra deg* (The Ocean Burns, 1966) and *Gada gredzeni* (Rings of Years, 1969), are sources of the poems here printed. Belševica is also an eminent translator of prose (Kipling, Hemingway, Axel Munthe).

TO BE ROOTS

To be roots. In that deepest earth where no ray
climbs down. Where light does not look in.
A treetop without birds. A leafless branch.
But the deepest spring strings out its capillary hairs
and must not break. The dirty work of roots
without respite. (Even winter's sleep is only seeming.)
Hoarding. Feeding. Giving drink. To be silent bond
tying life to bitter depravation. To give the joy of sun,
pronounce the strength of beauty through your
crippled unseen being to a white blossom.

To be roots. And not to envy blossoms.

(Inara Cedrins and Valters Nollendorfs)

DON'T BE OVERSURE

Don't be oversure of your cherry-tree.
Don't speak of roots. Of the fact that it's planted.
You see: in the vibrating wings of thrushes
a tree opens out branches. Berries tremble piercingly.
Birds don't call there. Leaves applaud
that easterner, who lifts the cherry in air.
And don't shake your head and say no,
the skies aren't a great Cherry Way,
but earth... the earth too will not say
against what it presses its grizzled nettles.
From green clouds a red rain falls.
You've a cherry in your garden? Don't be so sure.

(Inara Cedrins)

THAT WAS

That was a very polite fish...
when we were entangled in the sun-rays' net
which swung under the sharp wave edges,
the fish said: Glad to become acquainted
with a cold-blooded sister! How becoming to you
this blueness of face...
        I replied: Fish,
if I were blood sister to you
we'd not meet.
        It apologized
and turned tail to me.
        Take no offense...
I lie deep at the bottom of a forest
and above, spruces are green at wave crests,
striking up cloud froth occasionally in the wind.

(Inara Cedrins)

WILLOW-CATKINS

Willow-catkins threw off their narrow shoes,
danced in bare, fluffy feet.
All day the streamlets played reed-organs
through dwindling snowdrifts.
All day the woodpecker's celebratory pipe
fluted clear and gentle.
Breaking with fragile silver tongues,
the crumbling snow clinked.
Evening came, entangling in ice
and freezing the music.
Quietly beneath the sleeping willow-catkins
stole the hooves of shy deer.
Slipped on the willow-catkins' brown shoes,
went into the night's expanse.
Pointed tracks remained in the snow.

All in pairs.

(Inara Cedrins)

OLAFS STUMBRS (b. 1931)

A resident of Los Angeles, Stumbrs writes in both Latvian and English. He has written several volumes of poetry and has been a recipient of the Poetry Prize of the Zinaida Lazda Foundation.

CIGARS

Silence. Coffee. The dusky ceiling
takes our clouds, returns us no stars.
                                        We,
, three trumped face cards
dropped here by a careless dealer,
                                        sit:
: pinned mute through the mouth
to the pack of the couch
by this hour and our cigars,
                            Jon, Lee, I.
                            Lee, Jon, me.
    Unlike prayers,
                dreams,
                    sighs,
smoke rises to touch this private sky
playfully, easily—
                        Soon, anesthetized,
, far away iron the pain of whohowwhenwherewhy
all men reluctantly are,
we extract our cigars
unpinning the tongue and the soul, and
                        start in on the Primal Hen
with that cheery 3 AM dolefulness
                for having hatched such a universe:
:the heads of the rich
blind
inside rich purses,
our taxes too high, our joys too low
the important din
of our timid sinning,
perhaps
            blanking out some important astral "Hello"—
—Yes, and any vet with good reason could wait
to be anointed wealthy

as the Personal Doctor of the Personal Health
of any winning Presidential Candidate;
as for the state of real estate—;
and whatever happened to Baby Zen;
and have you heard of Cathy Greeneyes, you remember, when—
    Et cetera, and so on, and so frothily forth.
                       Almost forcefully
cutting myself off in midstrophe,
but proud to be my own sanity's guard,
I take another cigar
and again pin my head
            to the back of the sofa,
all three of us—sofa, cigar, me—
some staying, some rapidly turning green
with ennui.

ODE TO A COUNTRY CHILDHOOD

Everything changes.
Even the nose,
my classical nose,
my antique ruin
today is but a place to hang up glasses,
is but an instrument homely
for blowing syrupy tangos,
a vent
for getting the smoke out of my head.

But to find the sweetest strawberry patch in the clearing,
but to quiver from the odor of a cucumber freshly peeled,
but to smell a lollipop in the sticky palm of
        a friend—
that it can not.

My eyes see nothing.
In the morning eyelids yawn like sleepy lips.
The accustomed light of day
dries of all dew and soul from things.

Even the ears have grown deaf.
Surely, the wind still swishes about in yellowing
        oatfields,

surely, someone still sings inside a
    telephone pole,
surely, the pine tree still knows old
    squirrel stories—

Long ago I was a magician.
With an oakleaf for ship I cruised among green
    islands in a puddle of rain.
In the garden a wooden crate named GENUINE GOAT DUNG
    received me dear as home and sparkling like a castle.
And my cat, my good gray cat
never minded much changing into a dog, elephant, goose or a lion
who fiercely stalks among apple trees shaking its mane of fancy.

Precious world, what happened?

And sadly I sit watching my foot
    almost certain
that never never again will this simple shoe be
a roaring train which, throwing sparks,
fearlessly hurls itself into night.

(the Author)

A LIGHT, LATE SONG FOR SOLO VOICE WITHOUT PIANO
(Fragment)

And sometimes you are afraid.
You must flee, no matter where.
In the eyes of people and snapshots
gleam the sharp ends of bullets.
A piano key clicks
like the bolt of a rifle.
    You go for a walk.
    In the afternoon, Bach gives his children a whacking.
    Mozart worships fresh bread in the bakery window.
    From the park, you bring home
        a last year's maple leaf

and from it, from yourself,
    play the last autumn, the worries of frailness—
The evening comes with a red kerchief
(You are alone alone alone)
and ties it over your frightened eyes.
Then he sets you blind
in the fire of stars
    against the black
        wall of
            night.

(the Author)

A MARRIED ROMANTIC
(Three Poems)

1.
Love was holy hunger, and it kept my soul narrow.
It would fly straight up into the sun—a mad black free free arrow.
Now even blossoming maples tower above me.
(To fly in wind with white dandelion seed?)
Deeper into heavy trembling fright
quiet, steady-eyed serpents love me, love me—

What fasting must I do? How shall I pray? To whom?
I want my hunger back.

2.
Once I knew a girl in Salinas.
She looked her best at sunrise;
in the heavy gray-green of sunrise
she was my first pink flower.
All day
I wore her behind my hatband.
(Is this true memoir sorrow? That very adult pain?)

I should have kissed her harder.
I should have more to remember.

3.
This time is new to me.
The end of fairy tale: "And so, dear children,

they lived happily, happily..."
What now? Who speaks the next sentence?
I would have listened. Must I go to bed?
(Fat balding man starts adult story of middle-aged marriage.)

I should have no complaints:
my childhood was very long.

OJĀRS VĀCIETIS (b. 1933)

Vācietis is noted for his sharp pen and his outspoken criticism of restrictions on literature and the arts. He sees modern history as chaotic and undirected, and in his writing expresses his opposition to mechanization and materialism. His poetic work has been seen as close to the work both of Pasternak and of Yevtushenko. Our selection is from the collection *Gamma*, published in Riga in 1976.

BURNING LEAVES
"Manuscripts don't burn."
M. Bulgakov

What a musical score under the chestnut!
What fingers!

What novels
under the old linden!

What gnats about the light
under the oak chandelier!

What tiny, spark-like poem children
under the birch!

And they'll burn. And—ashes. And you, old woman, having gotten free of this rather messy work, you haven't gotten free. Only to you it's like just another year past in your life, but in eternity, honored, not only years don't disappear, but micron bits of minutes. I have to inform you dear, that, when spring is burned, or summer, fall — that is happening already in the next spring — and this one in the next; and I stand embalmed in this smoke

of yours, and you are a sorceress, and both of us reach into all of that which
is before us and all that will come after.

(Inara Cedrins)

SUNDAY

I go tiptoe
through my city—
through a dead orchestra.

That little wooden house,
dry and brown
like a violin,
of what can it think without strings?

Through two houses,
through drums
I squeeze by
and fear in vain—
they don't thunder me to the ground,
for in each is an open window.

The shattered hole of Sunday.

And then
I can no longer escape—
   in a shuffling gait
with empty eyes,
limp hands,
come the orchestra's musicians,
ours
and the children of noise.

At once they'll come up
from front and back
and ask to be allowed to smoke.

I won't scream.
They are children of noise.
But I am noise's father.

(Inara Cedrins)

HORSE DREAM

If one is judged by one's dreams,
prison is certain for me—
last night I led a horse
up onto St. Peter's tower
and said:
"Look, how I live!"

"It's nothing much,"
answered the horse,
"Though there are many tangles;
and, do you know what,
you'd better hobble me
for sometimes I start to fly.

The more you hobble
the less tangles will remain."

I grew thoughtful.
"No, but look, horse,
look, where I live!"

"I'm looking—"
said the horse,
"But stables don't determine a soul's color.
I am a simple animal
when I stand,
and a horse, when I plow."

...I hobbled the horse
and grazed it
by the planetarium
all night.

And all night
the stars bit me.

A couple passed,
and the boy said seriously to the girl:
"That's the fashion nowadays."

(Inara Cedrins)

57

OF NIGHTS

Of nights
some bear
wades through my oats.

I awaken
and listen: isn't that a pounding
someone
at my door
with a flintlock.

And the bear is undisturbed
in his oatful, sprouting sweetness
of eating.

They're all wrong
who say
that sprouting oats
aren't sweet.

As the last strumpet's breast
is sweet,
as a child—already fed—
still for that sweetness
shrieks.

Why
are we so unfair
to bears and oats,
bearing
this unfairness
like a top-hat?

They lie, who swear
that blossoming's difficult
and, if one breaks under that heaviness —
also difficult.

About my head
blossoming clouds
scuffle,
and in my oats
is a lucky bear,

and the field
through which he's waded
is, as much as that blossoming cloud,
lucky.

(Inara Cedrins)

IMANTS ZIEDONIS (b. 1933)

Ziedonis is today probably the best-known poet in Soviet Latvia, prolific
and established. His books of poetry have been known to sell out within
hours of publication. His work is impulsive, impatient, conscience-stirring;
he writes prose and translates, in addition to being a poet. These poems are
taken from *Kā svece deg* (How the Candle Burns, 1971).

HOW THE CANDLE BURNS

How the candle burns.
How beautifully the candle burns!
What a white light gleams, swinging.
And darkness flees. And again—light flees.
And God and the Devil barter for my soul.

But the candle burns.
How beautifully the candle burns!
And the wind runs up and frightens my flame.
Paraffin runs over the candle's edges
as someone already awaits my leaving.

But the candle burns.
How whitely the candle burns!
And darkness bows its full head in wonder:
from beginning to end it burns white,
till the wick slowly drowns in paraffin.

But something dims—
yet something lightly grows dimmer.
Still in my eyes the candle's light glimmers.
But before me stands the skies' great market,
where God and the Devil barter candle souls.

(Inara Cedrins)

INEVITABILITY

Chapped loaves of bread whisper through cracked lips.
Branches full of apples lean down from the heights.
People are like beets piled in a heap.
You too.

A strawflower burns in a vase, cut late in fall.
Beneath plump sacks the tired scales groan.
The churn turns slowly and whips up butter.
You too.

Two hands meet to lift a sack,
                two feet—to tread floor.
Two meet—to provide for a third life.
Inevitable. Inevitably
it happens each morning.
The desire to work.

(Inara Cedrins)

THAT IS HER MEMORIAL

That is her memorial.
A cast-iron pot.
An eaten empty cast-iron pot.
Come with your spoons,
beat spoons against your breasts
and say
that you didn't mean it like that.

Cross the spoons
and remember
that you are brothers at one bowl.
Brothers at one pot.
You, pilgrims with spoons—
bow your heads!
And pin up
all around her memorial
spoons.
Beautiful this blossom is not,
but honest.

All of us are going
to our mothers' memorial
with white spoon in hand.

(Inara Cedrins)

THAT WAS A BEAUTIFUL SUMMER

That was a beautiful summer.
We spoke only through flowers.

In May, we spoke through dandelions
and the words came with us like bees,
sticky with dandelion pollen,
yellowed.

There's joy too in speaking through pussywillows.

Hard to talk through lilacs—
in lilac nights words burn
and their ashes remain in the heart.

In July we spoke through poppies
and words remained in poppy heads.
Now, when I'm alone and no one is watching,
I rattle the poppy pods—
that's how I talk to you,
when I'm alone.

Then we spoke through nettles,
rather often we spoke through nettles.
Perhaps too much.
We probably didn't even need
            to talk through nettles.

Because in fall,
when we spoke through gladiolas,
one—through the usual rose,
the other—through carmine red;
already we no longer understood each other.

But later there were asters
so soon in their many colors:
you spoke through violet,
I—through yellow,
and I couldn't hear you any more.

And then everything had done blossoming.
There were left only
strawflowers.
Yesterday I put them in a winter window,
and strawflowers blossom there between two panes.

I sit by the window
alone, and speak
through strawflowers.
I say: "Soon there'll be iceflowers.
I won't talk through them—
not a word."

(Inara Cedrins)

MĀRIS ČAKLAIS (b. 1940)

Māris Čaklais has distinguished himself as being interested, in his poetry,
in the writing and ideas of earlier periods of Latvian culture. He has written
movingly on all ideological encroachments on the mind, whether religious
or political. He is noted also for his moving poetic portrayals of war.

...GIVE ME YOUR SILVER KNIFE

...give me your silver knife, right here
is that midnight tree (oh, grass,
how blackgreen you are in the moon's zenith!)—
let's peel the nightfruit together;
oh, if you haven't a silver knife, give
if only sharp nails, see
how blue the dew is on the skin;
oh, if you haven't, oh, if I have nothing else,
let's enter the midnight fruit with our lips;
with lips, with teeth, with fingers, with illusions

(somewhere beyond the hills shines hope's pale-blue razor-edge)
let's split it open and spill out, spill out
every last thing that we can possibly spill, and, fully
satiated and sticking to one another, let's awaken
when lackluster sleep is split by morning's spontaneous generation.

(Inara Cedrins)

DEDICATION

Seven years the boar's ham has been drying,
Seven years pain has glowered with sheathed nails.
Life alone as bread—wind on the exposed table,
measuring only once, cutting seven times.

But for millions of years the moon silvers fish,
for a million years circle the boar and pain.
Tear off a piece of ham, pain will tear it from you,
Let's toss this life a kopeck—for we do want to return.

(Inara Cedrins)

WOMAN IN MOURNING

Unobtrusive, with deep eyes, distant,
she came in like a black seaweed.

Languages froze on the spot.
Eyelids lowered like wings.

For a year now her son—a ghost has
drawn watercolors in heaven's garden.

What can one say—to the woman in black?
A word, before taking substance, turns to ash.

What has she to do—with us, who are alive?
Each one alive an unwilling tearer-open of wound.

Silence, like a spider, wove its web.
In the field trees drew on black garb.

Sadness, what did you want? What was it?
You sent no stand-in this time, revealed yourself...

(Inara Cedrins)

LITHUANIAN

INTRODUCTION TO LITHUANIAN POETRY

By Jonas Zdanys

Lithuanian poetry written since the end of World War II has developed within two distinct social and political spheres—the émigré and the Soviet—marked by clearly divergent modes of expression, thematic concerns, and formal structures.

One result of the incorporation of Lithuania into the Soviet Union in 1944 was a changed definition of acceptable literary standards. These new standards demanded that the literature produced be didactic in character, so that it might deal with the realities of a new society. Much of the poetry written in Soviet Lithuania in the forties and well into the fifties was limited in theme, colorless, and cliché-ridden.

A change in the literary landscape came in 1956, precipitated by the shakeups which followed Stalin's death. With political change and with the advent of the more tolerant Khrushchev, came both a loosening of political controls over writing and a change in literary directions. The monolith of "socialist realism" began to show some cracks. Poetry written since then—as the poems by the Soviet Lithuanian poets in our selection illustrate—shows variety in theme, a willingness to express individual rather than collective impressions, concerns, and experience. This writing is filled with historic association and allusion—including references both to traumas suffered before and after World War II and to Lithuania's mediaeval past—and, in the work of several of the important poets, a preoccupation with the drama of history. There are recent and contemporary cross-cultural allusions, revelings in fantasy often connected with Lithuanian folklore, and use of metaphor and symbolism. Soviet Lithuanian poetry, while still regulated, has in the past twenty-five years expanded in thematic scope, form, and structure, and has managed to enter the realm of personal, creative literature. Judita Vaičiunaite and Sigitas Geda are two particularly fine examples among the artists whose talents have matched the opportunities of this new freedom.

Émigré Lithuanian poetry has, for the most part, been more open to experiment and less constrained thematically than its Soviet counterpart has been. A number of Lithuanian poets—like some sixty thousand of their compatriots in all walks of life—fled the Soviet advance into their country and found themselves adrift in the cultural otherness of foreign lands. Out of that otherness, as if in response to it, grew a different kind of Lithuanian poetry. The poetry some émigrés wrote expressed a firm commitment to defining and maintaining a strong cultural identity. The trauma of displacement, especially in the work of poets of the older generation, nurtured a distinct and powerful nationalistic strain expressed through voices of protest and sorrow, of shaping a purified vision of Lithuania, the lost land, now extant only in the realm of preserved and refined memory.

Émigré poets of various literary "schools" continued and refined their work; others formed new groupings. Among the new groups, and perhaps the most influential voices among all Lithuanian émigré poets, is the Žeme (Earth) group. Of the five poets originally associated with Žeme three are still alive and two are represented in this selection: Henrikas Nagys and Alfonsas Nyka-Niliunas. Their poems, while different in formal and thematic specifics, combine to express a new poetic idiom, through it examining the experience of exile, the psychological and spiritual crises it poses, and the individual and collective regroupings it demands.

There are many other important émigré Lithuanian poets, but among those not affiliated with any school or group, Henrikas Radauskas (whose work was translated, among others, by Randall Jarrell) stands as an especially powerful and important voice. In his symbolist-surrealist poems he expresses the highest values of poetry as art, subordinating all other human experience to one—the aesthetic.

Readers interested in exploring more deeply a wide range of poetry of both spheres are referred to the quarterly journal *Lithuanus,* which frequently publishes translations of contemporary Lithuanian poetry, as well as to the anthologies *The Green Oak,* edited by Algirdas Landsbergis (New York: Voyages Press, 1962), and *Selected Post-War Lithuanian Poetry,* edited and translated by Jonas Zdanys (New York: Manyland Books, 1978).

HENRIKAS RADAUSKAS (1910-1970)

Henrikas Radauskas is perhaps the finest among the émigré Lithuanian poets. After studies in languages and literatures at Kaunas he made his living as a radio announcer and as editor of publications for the Lithuanian Ministry of Education. After coming to the United States in 1949 he spent a decade doing manual labor, and afterwards joined the staff of the Library of Congress, where he remained until his death. Poems from his four published volumes of poetry have been translated into English, Estonian, Finnish, German, Latvian, and Polish. His collected works appeared in Chicago in 1965.

HARBOR

At midday like a diamond
Locked up my eyes begin to fail.
The shore is charged with a fierce light:
A holiday of nails, broken glass, daggers, —
Who will give me a helping hand?

And steamer and locomotive
Sirens carve the rippled air,
And crabs and lobsters crawl
Between fishermen's stone hands,
And a crowd of screaming blacks
Pierces me as if with knives.

The shore is charged with a hot light.
Who will cover the fire of clouds,
Help me to wait for the cold light?
A holiday of lightning, flames, embers.

And the ocean rocks with boats
And glitters with crooked mirrors.

(Jonas Zdanys)

HOT DAY

A thin cypress scrapes the sky
And the hot day's perfume
Pours itself on the landscape's wounds.
Delicate needles pierce the heart.
Fainting she smiles and hears
Her own screams but cannot die,
The way you and I cannot.

She hears: a metal bird sings
In a glass tree, copper fruits trundle,
Tremble in the dizziness of dying.
Toward an old faun's golden foot
Swims soft music.

Having merged with the glass tree's invisible buds,
You don't care if tomorrow ever comes.

(Jonas Zdanys)

ARROW IN THE SKY

I am an arrow shot by a boy
At a white appletree near the green sea,
And a cloud of blossoms, like a swan,
Glittering descended into a wave,
And the boy stares but can't tell
The blossoms from the foam.

I am an arrow that a strong and young
Hunter shot at a flying
Eagle, but he didn't hit the bird
And wounded the big old sun
And blood poured on the whole evening
And the day died.

I am an arrow that a soldier not right
In the head, in a fort circled by enemies,
Shot at night toward mighty heaven
To ask for help, but, not finding God,
The arrow wandered among the cold stars
And was afraid to return.

(Jonas Zdanys)

SALESGIRL

Tears flow down the girl's face:
Loved, promised, doesn't write.
Summer can't be brought back
And she wrings her hands on the counter.

Castles collapsed, stories died.
The town is flooded with fall.

And in the store—ropes, horseshoes,
And nails, nails, nails, nails.

(Jonas Zdanys)

STAR, SUN, MOON

   Cabaret star Viola D'Amore (Violet Dam), who has never heard of
Vivaldi, walks out onto the stage clothed in smoke and scales—into a cloud
heated by applause and drinking, actions in a kerosene halo. Turning into
screams, which dim the mirror lights, she grabs onto the spotlight's fat
beam with both hands to keep from collapsing. The sun and the moon pray
in the heavens, worried about her future.

(Jonas Zdanys)

ALBINAS ŽUKAUSKAS (b. 1912)

The poet, short story writer, editor, and translator Albinas Žukauskas was
born in Poland. He studied law at Vilnius and journalism at Warsaw.
Among his collections of poetry are : *Ilgosios varsnos* (The Long Plowed
Traces, 1962), *Sunkus džiaugsmas* (Heavy Joy, 1969), *Atodangos* (The
Uncovering, 1971), and *Sagražos* (Reflexive Acts, 1973).

MIDWINTER TOURISTS

The memory persists, I cannot shake it,
Of those midwinter tourists, stretched out to rest
On the powdery snow of the beach
Near Klaipeda. The port city having just been liberated,
With canons booming a salute we gulp a hasty drink
And stalk through mines with the lieutenant colonel,
Slowly stepping in the tracks of tanks left on the snowy shore.
Such sunlight, ice-laden whirls of clouds, a burning frost!
Both fore and aft, as far as eyes can see,
On either side of the horizon, one long line
Of inert bodies. The lieutenant colonel jokingly decided—
They are tourists, gathered here from the new Europe,

Winter guests desiring rest who stumbled hither voluntarily or not
To spend the winter on a Lithuanian beach.
He joked and, for all I know, forgot, does not remember.
But I cannot forget them to this day
(Good, one must not, should not be allowed to forget!)—
On the frosty Lithuanian shore, petrified—resembling
A stony pavement brought by the ice-age once upon a time.
Next to the blocks of ice, by the water's edge, and even closer
To the dunes, a torn-up tract of ground. A wintry thunderclap—
A salute, resounds in tired bursts this side of the harbor.

A humming wave raises their garments—their uniforms.
Some turn their blackened chins toward the sun,
Others lie prone, still others on their sides, curled up
And yet somehow disheveled looking, some splashing,
Floating full length, tossed by the pounding billows up and down,
Revel in the expanse of the sun frozen blue into yellow ice,
In the brine and in the eternal, carefree leisure—
Contented, haughty, well-fed corpses of the Aryans!
Neither the past nor the present—the reality—excites them,
Even the future, that thousand-year existence of the chosen people,
                    they have secured for themselves
By reaching a grand goal, the conquest on the beach of Lithuania
Of patches of snow and sand—
The space of life!

I say, my dear lieutenant colonel, maybe you are right in thinking
That they are winter guests, midwinter tourists who blundered
Out of Germany onto the beach near Klaipeda,
Stretched out to bask in the icy heat. Well, I agree.
But even if that's true, these tourists are now meek and peaceful.
They won't rise in the evening to go home or to a tavern,
With arms around each other, or embracing
Sturdy, corpulent, pink-cheeked brides,
They will not holler, try to frighten, or gesticulate—

All of them were burned
By a different sun and lightning. Thus they will lie
Until the waves will lick away
Their flesh and bare the brown Aryan
Bones of a pure race, smooth them,
Then bury them in the sand so that there won't remain a trace
Of these here tourists...

There they repose, as far as the horizon and beyond,
Exposed to storms and sands, entrusted to decomposition,
The conquerors of the world, the icy winter guests.
Oh, you destroyed and then resurrected
shore of Lithuania! No longer will you have to fear
Anyone or anything!

(Irene Pogoželskyte Suboczewski)

EUGENIJUS MATUZEVIČIUS (b. 1917)

Eugenijus Matuzevičius, a poet, editor, and translator, was born in Russia.
He studied Lithuanian literature at the Universities of Kaunas and Vilnius,
and began publishing poetry in 1932. In addition to numerous works of
translation, the following works of poetry have appeared since the mid-
sixties: *Menesienos krantas* (The Moonlit Shore, 1965), *Vasarvidžio
tolumos* (Midsummer Distances, 1968), *Paukščiu takas* (The Bird Trail,
1970), *Žalios metu salos* (Green Islands of the Years, 1975), *Kol saule
musileis* (Until the Sun Sets, 1977).

THE WRITING IN THE FIRE

I don't know why
When I descry a flaming fire
On meadows, river banks, or near a wood,
I stop and do not venture any farther.
And later, right there by that fire,
I seem to run into a message
sent to me by an old friend.

I watch the flame in silence.
Inhale the bitter-tasting smoke...
And for some reason get the notion
That I've been sitting thus forever,
Next to this burning heap,
In the flames of which entangled
Seem reality and visions
That may never have existed.
And all of it,
That writing in the fire,

71

Gradually blends together
Into an ancient epic,
Which I now read
Like a message from an old friend.

Meanwhile in the dark, behind my back,
The night in chains is chewing on the grass.
And time, just like a gray-haired oldster,
Passes across the field, along the river bank,
Clanging copper bridles...

(Irene Pogoželskyte Suboczewski)

ALFONSAS NYKA-NILIUNAS (b. 1919)

Alfonsas Nyka-Niliunas studied Romance philology and philosophy at Kaunas. He came to the United States in 1949, and worked at the Library of Congress after having for a number of years held jobs as a manual laborer. He is the author of four verse collections, the most recent of them *Vyno stebuklas* (The Miracle of Wine, 1974). He has also rendered into Lithuanian the works of a number of English, French, German, and Italian poets, including Eliot's *The Waste Land* and Shakespeare's *Hamlet*.

THEOLOGY OF RAIN

A girl's footsteps
In the silence of the old face—
The inconsolable landscape
Where eyes
Paint the greenness black,
The erosion of being,
Unknown and never dried
Tears and the theologian
Rain's treatise to the roof.
Panta rhei.

(Jonas Zdanys)

WINTER LANDSCAPE

The city, etched in the black metal of sunset, is called Resurrection, city of my birth.

A large aureole of poverty and hate twirls it in a magnetic circle, and like a planet it revolves in the orbit of my mother's tightly shut eye.

The tree of life, dying of starvation, with congealed arms grasps for passing birds which slip free and fall dead to the ground. And the tree dies.

The sun—an anemone near the wrinkled brow—crawls past the violet well. Bloodstains on the earth-soiled threshold; snow on eyelashes; window without faces. The crazed door quietly sings the breasts and hair of the girl lying near the wall.

The city, etched in the black metal of sunset, is called Obliteration.

(Jonas Zdanys)

LiED

In the month of April, at night,
Despite the screeching of the sun,
We left for the north.
The strange talisman
On the girl's glittering arm
Grimly testified that longing
Would be hard for both of us.

Colors dissolved,
The numbers of birds increased,
And the sky more often
Reminded us of the blue of the dress
That danced around her knees,
Her unsmiling words,
Her pain in the gray beating
Of the seagulls' wings, and we
Understood then that she was dead.

(Jonas Zdanys)

MARCH

I still searched for your
Neglected silhouette in the dead window,
Your lips in the rain and your voice
In the loneliness of childhood's well; your steps'
Dark track in the sand,
Where the valley's blind echo laments
At the top of the black alder, near the path,
For the moon, robed for death by the wind.

(Jonas Zdanys)

BEFORE DAWN

A narrow sickle of moon.
A smell of sweet-flag and duckweed
In the reed kingdom.

Darkness changed to whitish silence.
Night ladled up a silver treasure.
The Milky Way
Descended to earth:
On it, returning from town,
Trying to sing something,
Walks my father,
A belfry under his arm.

The contours of stands emerge
On the stage of day. Soon will be heard
The first strains of the sun's overture.
Dreamed cities
Fade slowly in the windows
And in the well's cool waters gleam
The shepherd boy's sleep-filled eyes.

(Jonas Zdanys)

HENRIKAS NAGYS (b. 1920)

Henrikas Nagys studied Lithuanian and German ah Kaunas from 1941 to 1943, and Germanistics and art history at Innsbruck from 1945 to 1947. In 1949 he earned his doctorate at Innsbruck with a dissertation on the poetry of George Trakl. He has taught, after emigrating, at the University of Montréal, and has also worked as a commercial artist. Among his six poetry collections to date are *Melynas sniegas* (The Blue Snow, 1960), and *Broliai balti aitvarai* (Brothers, The White Goblins, 1969). He has also translated the works of American, German, and Latvian poets into Lithuanian.

HALLESCHES TOR: COMMUNION

Above the giant city dying in spastic convulsions
sinking into steam and smoke the setting sun bleeds and dies.
The hot ruins still scream in human voices
from the deep collapsed hells
of cellars.
In the evening's
thick red ember glow
the guards' metal faces, helmets, and bayonets.
The starving Badoglio prisoners,
dirty, half-deaf,
from hand to hand toss hot bricks and sing, muffled:
O mare mia...
Verfluchte Italiener!

* * * * * * * *

When at midnight the sirens howl again
and it rains
phosphorus, and mine explosions
rock the flaming sky—on Hallesches Tor subway
train's dirty platform we sit, twelve
strangers. The thirteenth—a pale
Badoglio prisoner—breaks apart and distributes
the last piece
of his moldy bread.
And later,
having rolled with his trembling and swollen
fingers a twist

of crumbling dry tobacco
he sends it around:—we each swallow smoke
(deep, eyes closed) like sweet wine.

(Jonas Zdanys)

POEM

I have begun to live on memories.
Evenings, when the fish
aren't biting,
I spread out yellowed photographs.
Read old letters.

At night
I try to find familiar stars.

Repeat another's lines
in a half-whisper.

Mornings
with a stalk of grass
I draw funny faces
on the water.

And feel sorry for
the spotted dead trout
lying next to me
on a bed of ferns.

(Jonas Zdanys)

POEM

I write no letters.
I dream:
in the time of blossoms
my sister
gathers red leaves.
My mother

76

picks winter apples
from leafless branches.
Each day I watch:
Canadian geese, honking sadly,
fly to distant
tropical lands.

In the evening
my sister and mother scoop
handfuls of lake water.

I write no letters.

(Jonas Zdanys)

LATERNA OBSCURA

Together we trace the child's face in the first snow.
Beneath wild raspberry branches my sister rocks her doll.
Last night workmen spread light snow on the frozen ground
and now tar the wooden bridge over the Bartuva.
The newborn snow is light as my sister's hair.

Through the cowering empty Samogitian village
the Cossacks ride, chopping the white mute moonlight
with their naked swords.

We trace our brother's face in the first snow.
The guard's epileptic daughter crumbles dry bread
into the coffin hole. Snow drifts over the peasant woman's
wax face and her plaited paper pillow.
Through the snowstorm echo the hoarse hymn and the breathless bells.

Through the soundless sleeping white Samogitian village
fly the Cossacks, chopping the blue winter moonlight
that shimmers in the trees with their long whips.

No one kissed you goodnight. No one wept with you
for your dead mother. No one came to bury your hanged father.
Your land was empty and naked. Your earth, a peasant's palm.
No one let you into the kingdom—grey garments fluttered
like long-forgotten funeral flags. Plague linens.

77

Through that poor Samogitian village fly the Cossacks,
carrying the chopping blue winter moonlight
on their long lances.

On a bright Sunday morning in the radiant land
workmen tar the wooden bridge over the Bartuva.
Deep beneath the ice the river flows slowly to the sea.
Under the raspberry branches sleeps my sister's snow-dusted doll.
Together we trace my sleeping brother's face in the blue snow.

(Jonas Zdanys)

Samogitian: Lithuanian Lowlands.

JANINA DEGUTYTE (b. 1928)

Born in Kaunas, Degutyte graduated from the University of Vilnius in 1955
and for several years worked as an educator and editor. She started
publishing in 1957. Her poetry collections include *Ant žemes delno* (On
Earth's Palm, 1963), *Siaures vasaros* (Northern Summers, 1966), *Pilnatis*
(Full Moon, 1967), *Melynos deltos* (The Blue Deltas, 1968), *Sviečia sniegas*
(Glowing Snow, 1970), *Tylos valandos* (Hours of Silence, 1978). In
addition she has also published translations, and numerous collections of
poetry for children.

ETUDE IN GLASS

The rainy fields are glazed with ice.
The earth has turned to glass. What if it breaks
under the cold and glassy sky?

This winter morning is as sharp and clear
as broken crystal
piercing the inside of a grasping hand.

Look,
underneath your skin
blood and desire flow,—
a cloud of red.

(M.G. Slavenas)

78

IN JAZZ RHYTHMS

This nameless anxiety
  grows beyond control
And we run away from home
  in thoughts or on trains or planes
And we spill into the streets
  and we fill hotels and cafés
Or set out on a raft alone—
  to confront the seas

Our eyes have gone blind
  from the dazzle of clever numbers and rocket glares
Our ears are deaf from loudspeaker voices
  and flutes in the spheres,—
Stunned and still
  we taste from a hand
  the sharp seed of the hemp.

Our spines are soft.—
  but we find it hard to hide behind walls of concrete and glass
  we slide up and down on the spiral stairs
And when we shall come face to face with our own self
  we'll wish to be like gods
  on the First Day of the world.

(M.G. Slavenas)

VYTAUTAS BLOŽĖ (b. 1930)

Poet, editor, and translator, Bložė was born in Lithuania and attended the Pedagogical Institute at Vilnius. Among his volumes of verse are *Nesudegantys miestai* (Unburnable Cities, 1964), *I š tylinčios žemes* (From the Silent Ground, 1966), *Žemes Geles* (Flowers of Earth, 1971), and *Poezija* (Poetry, 1974).

THE LEAVES HAVE YELLOWED

the leaves have yellowed like
veined faces: the wind

dashes them to the pavement: the coachman
like some ghost awaits the end of the banquet
when they'll swarm out, gay and intoxicated,
from whence there is no return: a felled tree
sways in the yard: its hard branches
seem to be fingers tinkling
on the heavy bronze gate: Aeolus'
rusting harp. Rhythmically
the passing years walk
through the house: the scream
of an infant and a small drop of milk
upon the lips: a warm
breast and tender
roots of hair. And dread. Where there
is nothing. Where only the wind
grabs the hand and does not release it
bareheaded Olaf, disembarking
on the sand-beaten shore, but
the rivers met him
with raised spears: and he fell
right here, where he is buried. Why this mass
of coal? Why the continuous return
of mounds of dross? The withered apple tree
blossomed and brought forth dry fruit. Whose
shadow over there is teetering and climbing
over the mossy roofs of threshing floors? And over
the damp ruins of castles? Suddenly
awake, he opens his eyes and yells
in an alien voice: behind these pine groves
where sunbaked deserts stretch, brothers
are selling Joseph to a traveling trader,
bartering him for sacks of hop
on which they then lie down to sleep. Now
it is easier. Glistening smoke
rises above settlements as though above
some ancient sacrificial altars: here
a sixteenth-century yard and chapel
shimmering in the rain, a stork's nest
and a woven blanket that has already
slipped under the horse's hoofs,
and some heraldic emblem like a muted voice,
while the draw-beam of the well
is swaying in the breeze,
its entire body trembling: tied to the pole

like a pendulum: tied to the well like a pendulum:
to a well in which the echo is profound
and I'm still crying in my sleep
while your patient mother stands
nearby with a candle and lights
the anvil, the peat and the mask
nailed to the wall

(Irene Pogoželskyte Suboczewski)

JUSTINAS MARCINKEVIČIUS (b. 1930)

A poet, novelist, dramatist, and translator, Marcinkevičius studied history
and philology at the University of Vilnius, specializing in Lithuanian
literature and graduating in 1954. A prolific and versatile writer, he has
published over ten books of poetry since 1955, some of the most recent
being *Sena abecele* (The Old Alphabet, 1969), *Poemos* (Poems, 1972), and
Šešios poemos (Six poems, 1973).

LANDSCAPE WITH APPARITION

A chapel—
like a kneeling mother
raises a silent grief heavenward.

The pain
is curled up on the gate,
the wail has hanged itself
on the draw-beam of the well.

Above the graves
black birds—
like funereal torches—
attempt but fail to burn.

Soldiers' helmets,
rusty helmets,
full of tears,
full of curses.

At sunset,
the clouds are torn apart
and reveal warming themselves
in front of the eternal fire,
perched on the white bench for dead souls,
our friends, our relatives and neighbors.

(Irene Pogoželskyte Suboczewski)

NEMUNAS

You are now full of the fog of autumn,
of silence and of cold. Your banks
are sloping toward the red western sky.

I know you are alive, you move as I do—
and I feel easier: you are like a hand
that I can either push away or clasp.

Woods get less dense. You move more slowly.
Look at us. Tell us about the land
that holds us by the hand along the bank.

Heaven joins in accompaniment as I speak:
give us a pagan creed and love,
Nemunas, now and always.

(Irene Pogoželskyte Suboczewski)

FATHER'S WINTER

The birds have all left
my father's tall trees.
Now only frozen stars
cling to the black branches.

Old farm tools stand stagnant:
plows, scythes, hands and hoes.
It seems there is nowhere to go,
nothing to make us wonder.

The wonders of Father's life
have passed unnoted and expected—
who will marvel at the water or grass,
or write down the spring or winter?

If you listen, at night
you can hear the deep and heavy sighs
of the senile, faithful Guernsey
passing up from the cattle-shed.

All thoughts, like snow, blend to one.

(Jonas Zdanys)

JUDITA VAIČIUNAITE (b. 1937)

Judita Vaičiunaite studied history and philology at Vilnius, specializing in Lithuanian language and literature and graduating in 1959. Her poems have been appearing since 1956; among her important poetry collections are the following: *Vetrunges* (Weather Vanes, 1966), *Po šiaures herbais* (Under Northern Coats of Arms, 1968), *Pakartojimai* (Repetitions, 1971), *Klajokle saule* (The Roving Sun, 1974), and *Neužmirštuoliu menesi* (In the Month of Forget-Me-Nots, 1977).

From *Four Portraits*

NAUSICAA

> "Hail, traveler, when you return
> to your own country, see that you
> do not forget me."

I've never kissed a man yet. My voice is like a wave.
    And my flesh has not been touched by male hands.
Yet I have hungered for one such as you.
    We were both dazed from exertion and surprise.
Both cursed and blessed be the ball we tossed around,
    having spread the linen out to dry,
that ancient golden morning when two funny mules had drawn
                my little cart...

* * * * * * * *

I am Nausicaä. I am descended from sea-faring ancestors.
There's something in me of a sinking ship fortuitously met.
My mother spins a thread of merriment, of purple wool.
And my father's open house is tall and generous.
So under ancient skies we're raising toasts both royally
and humbly
in honor of a lost and unexpected guest...
Why do you hide your tears under the mantle,
We are not trying to interrogate you—who are mighty,
mysterious and free...

I melt into the column against which I lean...
Let it remain a secret how I stood alone in the great hall,
For none will ever know what I was feeling then,
as I will not confess it even to myself:
I love you, Odysseus.

(Irene Pogoželskyte Suboczewski)

PENELOPE

"...with the name of Odysseus
in my heart I would go to my grave."

It is so hard for one who has known your love,—
for no one can replace you, no one is worthy of you.
Is it my fault that I am willing to wait for you for centuries,
to become deaf to the whistles and applause in the hall.
Is it my fault that the whole town is laughing, young men and
women slaves alike.
Is it my fault that I remain the dull and odd Penelope.
I want to be a glowing hearth to you across the barrier
of walls and years.
As pure as an idea,
I want to preserve for you the untouched amphora of pure water.
Let all generations of women feel the longing in my heart
when they hear the roar of the sea in a conch or see an empty room.
For I belong to those to whom one returns.
I am the faithful kind.
I became famous for all ages through patience and prudence—

I'll wipe the blood off your hands with my tresses and my lips.
At your knees I will burst into tears of joy.
I love you, Odysseus.

(Irene Pogoželskyte Suboczewski)

BOTTICELLI

Botticelli. Three candles. Soft lips suggesting a smile.
Swept off and carried away on a wave of light.
Silence ascends like fog. The ceiling expands.
The hardening wax covers the city in white.

I refuse to listen to the steady ticking of time.
I absorb from the painting the moisture of seafoam mist.
My rooms turn brittle. And the night exults
in the shell of your hand—this sunlit nakedness.

(M.G. Slavenas)

MUSEUM STREET

On the table—white dishes, bread and yellow apples.
And summer—beyond the opened window on the fifth floor.
Thunder and rain have quieted.
                              And the sun sketches
Itself round...
And a woman approaches the plaza—lighthaired and tall.
Drops of water on the roofs flashed for her.
Photographs of the holidays are ready.
These noisy, weary streets were laid with hot hands—
And the window,
          calling pigeons from towers
                              and sparrows,
Bread-feeders,
Rising like a high melody
                    above ghetto fires,
                              requiems and ashes...

(Jonas Zdanys)

TOMAS VENCLOVA (b. 1937)

Tomas Venclova was born in the Lithuanian coastal town of Klaipeda. He was allowed to publish only one collection of his poetry in Soviet Lithuania, *Kalbos ženklas* (The Sign of Speech, 1972). After this he fell into official disfavor, and was not allowed to join the Lithuanian Writers' Union. He became active in the Lithuanian dissident movement, and in 1977 was permitted to emigrate to the West (although without his family) as a result of petitions by Amnesty International, P.E.N., and his friends Czesław Miłosz, Arthur Miller, and Joseph Brodsky. Venclova's most recent publication in the West is his collected poems, 98 *eilerašciai* (98 Poems). He has been a Fellow of the Wilson Center, Smithsonian Institution, Washington, and has since 1985 taught Slavic Languages and Literatures at Yale University.

THE SHIELD OF ACHILLES
To Joseph Brodsky

I speak alone that on the nerves' taut screen
I shall see clearly now, as once you used to,
The key lying there beside the empty ashtray,
These railings by the chapels built of stone.
You weren't wrong: all's just as it is here.
For now. Even the scope of the imagination.
The same descent of kilometers to the shore-line
    Where still the sea

Hears both of us. Beneath the green-leafed roof
Gleam, almost as before, the heavy lampshades.
The different tempi that impel the clock's hands
Are far more dangerous than the bitter wave
Between us. Moving far in space's grip
You grow as distant as the Greeks, as strange as
The Medeans. In shame we've stayed, we others,
    On board this ship

Which is not safe, not even for the rats.
And if one looks well, then one realizes:
This is no ship, but brick walls, bright roofs, troubles,
A date that all too frequently comes round—
    In fact, maturity. This tutelage
Sinks into all our brains. Expanses,

Each day growing emptier, would have come to blind us,
     If by the verge

Where, vertical, the rain hovers and roves
A solemn vault of sound had not arisen,
Almost annihilated in this sudden summer,
But giving us the blessed manacles
That probably coincide with, fit the soul—
Exalt and burn, defining outlines, forming,
Because our heaven and our *terra firma*
     Are in voice, all.

Peace be to you. To you and me, both, peace.
Let it be dark. And let the seconds hurtle.
Through densest space, that dream of many layers,
I read each character your pen's released.
Whole cities disappear. In nature's stead,
A white shield, counterweight to non-existence.
In its engraving both our different eras
     Lie double-etched

(Were there but happiness and strength enough!)
As though in water. Or, put more precisely,
As though in emptiness. Waves beat the beach-head,
Disintegrate the mobile sketch. The squares
Of windows gleam with blackness. Late in dreams
The heated air seeps slowly through the glass panes.
Beyond the towers, a motor faintly rasping,
     And into me

Roll day and hours. You see, between each chime,
The bell's blind swing inside its belfry.
Till the foundations answer its peal dully
There flows an endless interval of time.
The portals quiver, tautened by the beat,
And archway signals out to neighboring archway,
And souls and continents call out to one another
     In living night.

A dirty gloom enshrouds the sails, and sticks.
The sodden quay exhales a pungent vapor.
You see Thermopylae, having seen Troy earlier—
The shield is given to you. You are a rock.
The pillars set above this permanence

87

Impact the wind with their scintillant metal,
Although the rock, too, stands near sham and swindle
    And wordlessness.

Entrusting to each one of us our fates
You cross now to the level of remembrance.
But every moment that exists, exists twice.
We are accompanied by a double light
Inside the ring that days, nights tighten more.
Low tide. On sand the ebb's pools glisten.
Boat, stone don't yet look different on the coastline,
    The empty shore.

(David McDuff)

NIGHT DESCENDED ON US WITH A CHILL...

Night descended on us with a chill.
The low-roofed and soot-blackened archways
Looked onto ten stations as well
As ten parks, or more, sunk in November—
That settlement, circle, or zone
Where on the blind brickhouse wanders
A moving one hundred watt beam:
In the labyrinth, mentor and escort.

Temporarily we make our home
In the kingdom of Ariadne and Minos.
Because of the fog and the gloom
For hours not a single plane takes off.
Every day again all trains are jammed—
How much space, how much air and unhappiness.
So those prisoners who returned home
Sometimes longed for the eye of the cell-guard.

Like a debt repaid to the void
Some familiar places stood opened:
I repeated inside my head: "bus,
University, after monument, island."
I said: "Tomorrow I'll go,
I'll go or at least I'll try to."
And along the hither world's brow
My soul hurried on into limbo.

Old addresses grew suddenly near.
The alphabet changed form and meaning.
Voices grew faint and dead, I could hear,
Unable to find us two in either
This house's locked up, empty cell
Where the paintings don't recognize me,
Or in dreams, or in heaven, or even
Dante's second circle of hell.

Thus time is stopped; to be exact,
One cases existence gradually.
It's just that each year, in effect,
You hear the 'phone ring more remotely,
And memory, day after day,
Shifts diameter like a compass
Till the past's a straight line on the page,
First pretending to turn into distance.

What you hear and see, I can't tell,
In reality chipped from reality.
The paved banks of Acheron withheld
The unfelt swell. Each nullity
Is separate, all on its own.
And the world lives its life without us.
There exist, in the end, alone
Dead silence and the nine muses.

Where the capital slowly revolves,
And the snow's games make us weary,
Where the fog hides all objects' selves,
Thank God for the dictionary.
In the kingdom where a friend's hand
Will never hurry to help you,
The void or the supreme power
Sends the angel—rhythm and language.

I ask not one moment's respite,
Neither death, nor forgiveness for sinning
—Only leave the primordial drumming
Over stone and the ice of the night.

(David McDuff)

IN MEMORY OF THE POET. VARIANT.

Have you come back, then, to the promised city,
Its plan, its skeleton and carbon copy?
The storm has sailed away the Admiralty,
The geometrical paints in the city squares
Have faded.
          Power is disconnected
By shadow born inside an icy spectrum
And not far from Izmailovsky Prospekt loom
Steam locomotives rusting in the air.

The same tramcar, the same old threadbare topcoat...
A piece of paper, lifted by the asphalt,
And all's gripped by a nineteenth-century cold
That chokes the station up.
              The droning sky
Snaps shut. The decades fall and die together.
Like storms the cities pass, a violent weather,
Like gifts, our gestures mirror one another.
A man, though, isn't born a second time.

He melts into the February morning
That hugs this Rome, lying inert and northern,
Into another space, choosing a rhythm
Drawn close in time to fit the hour of snow.
The frozen cavern of the she-wolf calls him,
The mud, the madhouse and the walls of prison:
St Petersburg, the black familiar city
That people spoke about so long ago.

Not born again are harmony and measure,
The crackling boards and the hot acrid odor
Of the hearth that time both lit and nurtured
—but after all, it is a timeless hearth
And optics that are destiny's last judges,
Its essence being luck, coincidences
And sometimes nothing more than chance encounters
And the continuance of eternal forms.

A crack in what is real, not a reflection,
An island grown into the foaming ocean
In place of paradise's revelation
Shells off the husk of living speech and words

And in the clouds flooding above the boat's horn
Doves spin a wheel that still, gigantic, floats on
Not daring quite to sever Ararat from
The ordinary blossoming hill's breast.

Let's leave the shore. And sail. It's time now.
Stones crack asunder and the falsehoods ooze out.
But art remains, a solitary witness,
It has survived the deep midwinter nights.
The grasses tame the ice-tracts with their blessing,
The rivers' mouths find dark seas, rediscovering,
And here a weightless word goes echoing
Almost as void of meaning as a death.

(David McDuff)

VILLANELLE

Break off. The phrase disintegrates and dies.
The roof-line coincides with crack of dawn.
Snow speaks, the fire re-echoes what it says.

The lead weight gives the floor a gentle graze,
By hours the circle slows its sway and swing.
Break off. The phrase disintegrates and dies.

Where once the world was, now a drawing shines,
Reflected in the mirror's empty moon.
Snow speaks, the fire re-echoes what it says.

The convict files back to his cell's recess.
Up, skywards trails the barbed wire of the zone.
Break off. The phrase disintegrates and dies.

A fraction out of time, a shard of space
Grips like a sphere your body and my own.
Snow speaks, the fire re-echoes what it says.

To each face presses all that vanishes.
The bed's head has no angel guardian.
Break off. The phrase disintegrates and dies.
Snow speaks, the fire re-echoes what it says.

(David McDuff)

DIALOGUE IN WINTER

Go in, enter this landscape. It's still dark.
Beyond the sand-dunes drones the empty roadway.
The continent fights hard against the seawaves—
Invisible, but brimming full of talk.
A traveller or an angel left behind
This track grown taut with radiant light, snow-dusted;
Recalling barenness, Antarctic land,
Against the windows' black: the shore's reflection.

The deep sea foams still, has not frozen yet.
The sands have blown more than just one mile inland.
The quay is quite distinct here; here, it falters
As the great cavity of winter grows and spreads.
There are not letters and no telegrams,
Just photographs. The radio isn't working.
As if a candle had imposed its seal and stamp
Upon this time of danger with its melting.

How damp the air, how resonant the stone,
How powerful the beam of daybreak's roentgen!
Screwing up your eyes, you see walls grow transparent,
The church tower, and the figure of a man.
Only the foggy contours of the trees
Show on a white background. And through the bark's rind
Although your eyes are closed you almost see
The last ring coiled up, tight, resistant.

"That custom's tiring for the eyes, besides
After an hour it's easy to make errors."
"Prophetic truth doesn't meet the situation."
The hoar-frost covered axis slips and slides,
At the horizon's far extremity
It seems, where ships sail black, and each sound stiffens,
That in the sluggish sky above the sea
The planets Jupiter and Mars glint, flicker.

The emptiness spreads to the Atlantic coast.
The fields are bare. They stand like empty hallways.
And February hides under January's layers.
The plains cower from the wind's drenched, drenching blast.
By the lagoon, hills show themselves quite bare.
In hollows, melting snowdrifts seep and dwindle

And blacken. "There, what is that, over there?"
"Again, just estuaries, bays and harbors."

Beneath the heavy trawling net of clouds
Squares glitter, narrow, bright, like fishes.
"Do you remember what it was the stars said?"
"This age rolls into being without signs,
By force of facts alone." "The gravity of death
Enchains each person, every plant and object,
But grains and offerings sprout new breath
And that is why I think not everything is ended."

"Where is the evidence?" And who the one
That separates the truth from lies and falsehood?"
"Perhaps we're here alone in all the world."
"It seems to me, too, there is you alone."
"And the third speaker? Are you saying then
That no one overhears our conversation?"
"That there are heaven, fields, snow covering them.
Sometimes the voice outlives the heart's ambition."

Noon makes the trees seem darker than they were.
Now, in broad daylight, you are conscious only
Of small things, scratched an hour ago from nothing,
Which stand in place of names, instead of words.
A splinter broken from an ice-cube tray.
A skeleton of boughs. A crumbling brick house,
Near where the road bends. Later in the day
On this sea-coast and on the other—stillness.

(David McDuff)

SIGITAS GEDA (b. 1943)

Sigitas Geda, who studied history and philology at the University of
Vilnius, is perhaps the finest poet of the new generation writing in
Lithuania, and among the best of the Baltic writers who have emerged since
World War II. His collections of poetry include *Pedos* (Footprints, 1966),
Strazdas (Thrush, 1967), and *26 rudens ir vasaros giesmes* (26 Songs of
Autumn and Summer, 1973). He has also published several important
cycles of poetry in journals, including poems in honor of François Villon
and Pablo Picasso.

I WALKED OUT INTO LITHUANIA

I walked out into Lithuania.
There were birds, women and wind.

All the cows walked towards day,
And a large—brown—meadowlark
Fluttered and flapped in the wind.

Long wings and whisking rivers,
And all the women—butterflies,
My eyes grew over with grass,

And from the very depths of the grass
Animals turned red.

Was I there or wasn't I,
Or did I dream a clay dream?

(Jonas Zdanys)

REPENTANCE: THE DEVIL'S BLOSSOM

And who now will wake the dead:
the endless ocean waters
poured over the dark sword, rusts
and reed leaves invite you there,
bright fire, and all around the sinister
winds of ravines, and the confused dog
barks by your face, and the wings of the angel
stop you from walking the road...

And who now will wake the dead:
I found the sweetbrier by the road
and said: within this halo watches
a thorny god, I give it to you,
bright fire, I spread myself wide...midsummers
pant in the terrible
glare of the grass, when I, kneeling,
with my own hand murder myself...

And who now will wake the dead,—
I drank the dark blood, the moon
shone unnoticed, I might not have murdered,
but nettles grow green in the sun, and here
begins the pit of the useless,
and fires are so bright for the one who walks
that the devil's blossom and the waned moon split...

(Jonas Zdanys)

GOD'S FAMILY

Once in the universe ripened
God's small family: a wife
and a small boy, who looked
at the great blue evening
with dark eyes,
and a husband—a brave musician,
a pleasant singer from the circus,
who loved to drink wine
the color of smoky grasses.

Once in the universe ripened
God's small family:
on wayworn legs the boy
carries an ant on his
palm toward the elderberry bush
swaying in the night...
The dark-eyed woman, alas, didn't know
why it was all necessary
and knitted far into the night.

Once in the universe ripened
God's small family,
and there is no one to tell now
what awaits them, what will
still be...Toward the dusty
elderberry falls the reddening
blossom of the stars,
and paled lips articulate
a single word: death...

(Jonas Zdanys)

SEBASTIAN'S LAMENT, 1943

The nightingale voices of memory
led me, it was not
terrible tonight as I walked home,
and I said that I feared nothing,
the colors of red...

And then something
near the wasteland wrenched my arms,
I wanted to cry out, but in the stillness of fame
the devil flew across the sky, in the distance
fired arrows...

And the dark throng
gathered them and pierced me
until my blood glistened in the sand,
where will I find the grass to lie in,
the colors of red...

The void of the sands,
the blood of slaughterhouses, carrying her head
my mother passed me by, heaven,
I gathered your white blossoms,
wanted to weave them...

Alas, totally alone
the mute bodies wail in the water,
in the fire and the wind...

(Jonas Zdanys)

STEEP EYES OF WOODEN GODS

Steep eyes of wooden gods—
Are they not my
Not your eyes?

How close you are
My forefathers!

From where does wind tear the tracks
Of a hundred years from the roof?

Tracks, tracks,
Tracks and tracks—
Men
And their gods
Stopped.

Such crows,
Starlings,
Lived in the
Fifth and the twelfth centuries.
An old man hammered a nest for the brown one,
The brown one's voice is brown and eternal.

How close you are
My forefathers!

They herded cows
And saddled horses,
Planted children
And peas.

Drink.
Pray.
Pray
Drink—
Gathered within me as if in the ground.

(Jonas Zdanys)

VIOLETA PALČINSKAITE (b. 1943)

Palčinskaite studied history and philology at the University of Vilnius and made her debut as a poet in 1958. Much of her work is in literature for children. Her works of poetry include *Žeme kele žole* (The Earth Raised the Grass, 1961), *Akmenys žydi* (The Rocks Are Blooming, 1963), *Aikštes* (Squares, 1965), and *Kreidos bokštai* (Towers of Chalk, 1969)

SILENCE

1
The castles of Čiurlionis—rise again
In silence—in an alien city.
An ash-gray fog is spreading mutely
And freezes into immobility within us.

The meaning lies in artifacts, in clothing.
(We're used to them since childhood.)
And miniature comets are swaying,
Poised on the blade of a knife.

2
...And on the plate are ripe large brown nuts.
Do stay. I know how to prolong the hours.
Because for seven long years I walked to you from afar,
When all could see and when no one was looking.
I carried this light—unusual, pure, even.
I exist. I am sure. Do not disappear.

                                             My skin is torn,

When simply breathing can dissolve the twilight...
The brown nutshell hides a hollow heart.

3
I am rocking the silence in my arms
Like a sleeping infant.
Where are the tree hollows that always offer safety
To those who lie down under blankets of leaves.

A yellow shadow will continue growing
Next to the thick beam of a spruce.
And golden orioles will call for rain.
If they could only find another earth...

4

To Harpocratus—the god of silence—
Let us dedicate this secret. After all,
Every day on roadsides and construction sites
Living shells know how to turn into stone.

Sometimes physical life is exalted,—
Its mute truths will be found in experience.
You're as sure as an open incision,—
That blends life and death.

5

I will immure the wine—the red, the bitter,
Like a vintner's toil—sacred and true.
Later, only once in the future,—
When fingerprints appear on black glass,
So that springs of the century do not stop flowing—
Then you will suddenly burst in like a festival
Through the solemn silence and narrow heavenly vault
Into my harmless soiled cellar.

(Irene Pogoželskyte Suboczewski)

POLISH

INTRODUCTION TO POLISH POETRY

When in the fall of 1980 Czesław Miłosz accepted the Nobel Prize for Literature he called attention once again to the achievements of Polish letters; equally important, he sounded anew that keynote of conscience that he had let ring the decade before: "The act of writing a poem is an act of faith; yet if the screams of the tortured are audible in the poet's room, is not his activity an offense to human suffering?" (*The History of Polish Literature* [New York: Macmillan, 1969], p. 458). Is poetry still possible after Auschwitz? The answer offered by the best Polish poets agrees with what good poets think and feel elsewhere. Yes, it is, but we must redefine the task of the poet. Poems must continue to be written, but not as a pastime. Poetry worthy of the name is a question of life and death.

Poland is by far the largest of East European countries—almost the size of the two Germanies combined—and it should not be surprising to find that its ongoing literary activity is spread out over four or five of the important metropolitan centers: Warsaw, Lódz, Cracow, Poznań, Wrocław. Partly for this reason poetry, across the work of the important groups (e.g., "Agora" at Wrocław) and the members of three active generations of poets, displays a variety and a range of inventiveness, a great tree grown from the mustard seed of war and suffering. It would be tempting to categorize, and to divide the fourteen poets by whom poems can appear below (along with the at least fourteen others of whose work we are unable to print any examples) by school, "group," or by the ever-convenient chronological bias. None of these approaches would give a true picture of the phenomenon of constant movement, of ever-present retrospect and renewal. Neither would any critical method that merely discriminated while it did not also show how poets reach out and grasp one another by the hand, do justice to subtly changing poetic textures and to combinations of interests.

Foremost among those to whom poetry is an ethical act, with aims that poetic language must serve, is Tadeusz Różewicz; rejection of the traditional devices of versification—prosody and metrics, metaphor and symbol—enables him to build an oeuvre of poetry characterized by utter "nakedness," plainness, naturalness. In the thematic hemisphere of ethical concerns Różewicz's art is not one given to forgetting war and holocaust, and his early postwar volume, *Niepokój* (Anxiety, 1947), identifies his voice as one of those that will crucially matter. Among poets appearing

101

here, the older Czesław Miłosz and the slightly younger Zbigniew Herbert are close to Różewicz in spirit, although both have a decidedly expansive, classical outlook. Especially Herbert's diction is characterized by a tendency to reexamine the historic past and to invent alternate models for the solutions of the great tragedians (as in "Elegy of Fortinbras" or "The Sacrifice of Iphigenia"). Two further important variants on the "naked voice" in poetry are the experimentalists Miron Białoszewski and Tymoteusz Karpowicz. Białoszewski's treatment of tubing and scaffolding, which strongly reminds one of Jerzy Harasymowicz's poetry of umbrellas, as well as (outside Polish writing) of Vasko Popa's cycles involving totemic repetition, seems to ask the crucial question whether it is possible to transform the semantic organism of language into its formal moment. Beyond the earnestness that is undoubtedly there, Białoszewski is not indifferent to the possibilities of playfulness and humor; in this he is rather different from Karpowicz who, while by no means humorless, places the center of gravity of the poetic utterance not on subject matter but rather on the fascinating poetic of the logic of language. So much depends in Karpowicz's *Reversed Light* upon typographical presentation that the artist qua artist can never manage to vanish, however temporarily, behind the stage set of his artifact; a Brechtian alienation device is present in the wings.

How many possibilities exist for addressing the earnest business of poetry as a real moral act is clearly shown in the work of poets who do not exclude from their world the elements of innocence and delight in natural beauty. In his very helpful introduction to his translated selection of Harasymowicz's verse entitled *Planting Beeches* Victor Contoski writes: "Harasymowicz delights in juxtaposing artifacts of civilization and elements of the natural world—pianos and frogs, inkstands and rivers— and in tracing civilization back to its beginnings" (New York: New Rivers Press, 1975, p. 7). No less do Jan Bolesław Ożóg and Tadeusz Nowak, poets seventeen years apart, delight in images of village life, the demotic, the natural. Both Nowak's psalms and Ożóg's "Song of the Field" return to a lyricism that is a distinct respite after the poetic fundamentalism of the school of Różewicz (if we may speak of a "school"), and which makes us thankful for authentic variety and seriousness both.

If there is any "artificial" boundary line that we could draw on Poland's imposing poetic map, it is the utterly natural one of poetry by men versus that by women. Because differences as well as (not always elective) affinities both delight and instruct, to discriminate here is to help overcome discrimination. Anna Świrszczyńska's highly personal voice seems well represented on these pages; one risks a guess that its owner was distinguished long before she became senior. And she is far more the "woman poet" in the traditional sense than either Wisława Szymborska or Ewa Lipska. Szymborska's sheer range is, in fact, deeply impressive, from the language-conscious "The Joy of Writing" to the stark generalism of "Preparing a Curriculum vitae" and the dialogue form of "Words." With Karpowicz Szymborska shares the true poet's conscious love for and commitment to language; with Różewicz she holds in common her view of contemporary life as an age of man in which recapturing humanity and hope are acts of uncompromising heroism. Perhaps the playfulness and yet deep lyricism of Ewa Lipska, born in the critical and hopeful year 1945, can best round out our image of a poetic scene in which all agree on an unquestionable goal for poetry, while few choose to follow the same or even similar means for attaining it. Iconic for this scale of differences in Lipska's own method are poems like "That Moment" and "When Our Enemies Fall Asleep." That pact of non-aggression will yet be signed.

Among poets whom to our regret we could not include, at least mention should be made of Władysław Broniewski (1897-1962), Stanisław Swen Czachorowski (b. 1920), Stanisław Grochowiak (b. 1934), Urszula Kozioł (b. 1935), and Adam Zagajewski (b. 1945).

JAROSŁAW IWASZKIEWICZ (1894-1980)

Iwaszkiewicz was born in the Ukraine and died in Warsaw. He was an enormously versatile and prolific writer, who worked in poetry, the novel, the short story, the essay, the drama, and translated from the English, Danish, French, Spanish, and Italian. He was one of the founders of the important prewar group of poets named Skamander, but after World War II he came to be appreciated more for his prose fiction. From 1954 on he served as editor-in-chief of the literary monthly *Twórczość* (Creative Work), a distinguished journal that maintained consistently high standards throughout the entire postwar era. In 1970 Iwaszkiewicz was awarded the Lenin Peace Prize. His prose fiction shows him to be that rare twentieth-century Polish writer who continues in the traditions of Turgenev and Bunin. His poetry, for the most part, is highly formal; but the two poems offered here, taken from one of his last volumes, demonstrate that he could also write in a more relaxed and more recognizably "modern" manner.

THE POPE IN ANCONA
           To Wojciech Karpiński

In the crystal sky
of the fresco
the gerfalcon fell on the pheasant

The ailing Pope has come to Ancona
The Pope wants to move against the Turk
he has implored all the princes
to render him assistance

The Pope says:
"not yet is he entrenched in Byzantium
he can still be smoked out"

But he cannot be heard
for he can now barely speak

That is why he became Pope
(which was not easy for a poor scribe)
to drive out the Turk

But no one has come to Ancona
and the Pope is dying
and the Doge regrets
having assembled a fleet
for the fleet will not sail
against the Turk

And one will not see how the gerfalcon
tears the pheasant to pieces

(Magnus J. Kryński and Robert A. Maguire)

HOW DOES THE NEGATIVE LOOK...

How does the negative of a color photo look?
What purpose do television antennas serve?
What is the basic principle behind mathematical computers?
What must a man know
who is flying to the moon?

What is the circumference of
the cupola of St. Peter's equal to?
Why the structure of the crystal?
What do Piaget, Ricoeur, Adorno, Starobiński say?
Why doesn't Lévi-Straus like Malinowski?
Will Jakobson's researches change the structure of *Pan Tadeusz?*
What is the meaning of the cry: Galilaee vicisti?

Why did the Golem enter my house?
Why is Mr. Hyde sitting at my desk?
Why is the devil's funeral being held beneath my window?

All this must be destroyed.
Let there be only a quiet beach and a complete absence

of questions

no questions

(Magnus J. Kryński and Robert A. Maguire)

ANNA ŚWIRSZCZYŃSKA (1909–1984)

Świrszczyńska was born in Warsaw, and since World War II has lived in Cracow. She has written plays and children's books, but has been known as a poet since her well-received debut in 1930. She has moved from a highly stylized manner to a great simplicity and straightforwardness. Two of her most recent collections of poems have been best sellers in Poland: *Jestem baba* (I Am Female, 1972) was perhaps the first book in postwar Polish literature to strike a defiantly feminist note; *Budowałam barykadę* (Building the Barricade, 1974), a cycle of a hundred poems, was one of the first treatments in verse of the Warsaw Uprising. It has been rendered into English by Magnus J. Kryński and Robert A. Maguire, and was published in a dual-language edition by Wydawnictwo Literackie of Cracow in 1979.

WHEN WE WAKE UP IN THE MORNING

How good that it happened
how good that it's already over.
Thank you, sweetheart,
For these two joys.

Now
my body is light and my soul pure.
Dry-cleaned of lust
I have a violent craving,
like an asthmatic for air,
for tasks that are hard,
for hard human work.

My head wants to work and my hands,
I have been created for work,
not for pleasure.
I am strong, I can carry
burdens, as do the strong.

(Magnus J. Kryński and Robert A. Maguire)

THE FIRST MADRIGAL

That night of love was pure
like an ancient musical instrument
and the air around it.

It was rich
like a coronation ceremony.
It was carnal like the belly of a woman in labor,
and spiritual
like a number.

It was only a moment of life,
and wished to become a conclusion drawn from life.
Dying
it wished to know the principle of the world.

That night of love
had ambitions.

(Magnus J. Kryński and Robert A. Maguire)

THE GREATEST LOVE

She is sixty. She is having
the greatest love of her life.

She walks arm in arm with her sweetheart,
the wind tousles their gray hair.
Her sweetheart says:
"Every hair of your head is like a pearl."

Her children say:
"Crazy old woman."

(Magnus J. Kryński and Robert A. Maguire)

CZESŁAW MIŁOSZ (b. 1911)

Born in Lithuania, Miłosz studied law at the University of Vilnius, where he was active in the literary group Zagary in the thirties. After World War II he entered the diplomatic service; since 1951 he has been living in the West, and has been professor of Slavic Languages at the University of California, Berkeley, since 1961. In 1978 he received the Neustadt International Prize, and in 1980 he became the recipient of the Nobel Prize for Literature. A poet with a powerful and expressive range, mostly in freer forms, Miłosz has published numerous collections of his verse in Polish, as well as works of critical prose. He is author of a seminal *History of Polish Literature* (Macmillan, 1969). Miłosz's verse has been widely translated; the Ecco Press in New York has published two translated selections of his verse, including *Bells in Winter* (1978).

SCREENS WILL BE SET THERE

Screens will be set there
And our life will be seen from beginning to end
With everything we managed to forget, it seemed for good,
And with the costumes of the time that would be only ridiculous and pitiful
If it were not we who wore them, not knowing any others.
Armageddon of men and women. It is futile to shout I loved them,
Each one seemed a greedy child longing for caresses.
I liked beaches, pools and clinics,
There they were bone of my bone, flesh of my flesh.
I pitied them, and myself, but this is no defense.
Every word, every thought is gone: moving a glass,
Turning the head, fingers unbuttoning a dress, clowning,
A cheating gesture, contemplation of clouds,
Killing for convenience. Only this.
What if they depart with the jingling of bells
Around their ankles, if slowly they enter the flames

That took away both them and me? Bite your fingers, if you have them,
And look again at what once was—from the beginning to the end.

—written in 1964, in Berkeley

(John and Bogdana Carpenter)

ON THE BOOK

We lived in strange, hostile, marvellous times,
bullets sang above our heads
and years no less threatening than tearing shrapnel
taught greatness to those who did not see
war. In the fire of the dryly flaming weeks
we worked hard and were hungry
for bread, for unearthly miracles appearing on earth
and often, unable to sleep, suddenly saddened
we looked through the windows if over the blue night
flocks of zeppelins were not flowing in again,
if a new signal did not explode to the continents
and we looked in the mirror if a stigma did not grow
on the forehead as a sign we were already condemned.
In those times it was not enough to lament
with pure words the eternal pathos of the world,
it was an epoch of storm, the day of the apocalypse,
old nations were destroyed, capitals turning
like a spindle, drunk, under the foaming sky.
Where is the place for you in this age of tumult,
wise, quiet book, alloy of the elements
reconciled for eternity by the sight of the artist?
Never again from your pages will a foggy evening
glisten for us on the quiet waters as in Conrad's prose,
or the sky break into speech with a Faustian choir,
and the long forgotten song of Hafiz will not touch
our forehead with its coolness, will not rock our heads,
nor will Norwid reveal to us the harsh laws
of history hidden in a red whirlwind of dust.
Anxious, blind, and faithful to our epoch
we are going somewhere far away, above us October
murmurs with a leaf as the other one flapped with a flag.
The laurel is not for us, aware of the punishment
that time allots to those who loved

temporality, deafened by the din of metals.
Thus we were marked to create a fame—nameless,
like a farewell shout of those departing—into darkness.

—written in 1933, in Wilno

(John and Bogdana Carpenter)

JAN BOLESŁAW OŻÓG (1913–1991)

Like Nowak, Ożóg is also the child of a peasant family, and almost all of his creative work has dealt with the tragedy of the fate of the peasant: the conflict with nature, the transience of old village life, and the fate of Christian mythology in the face of modern psychology. He has published over twenty volumes of verse, as well as works of literary theory and criticism. To our best knowledge he has thus far been untranslated in the United States.

VILLAGE FOR A WEDDING

Sky like pigeon's little belly.

Sky like goldfinch egg,
sky like starling's tune
greenishblue.

But fields like sundrenched sea
where roe-deer bound through oats
like fish through sea of rye.

And the village distant from a hummock
like a chain on a bicycle.

Iris in golden glimmer like pilot-flame.

But trees deeprooted like ponds
of green broth.
But grass like prayers
from lips of decaying willows.

And the village distant from a hummock
harrows gouged by nails
instead of stakes.

Lady bugs slide off lindens,
cockroaches to pick in a kerchief.

In barns peasant gears groan
cranking round the chaffcutter,
and frightened wasps play
like a heathen church at high mass.

Red beaks of carrots
circle higher than storks.

And the power station beyond the village
limping on crutches
like bent herdsman on crook.

Here the morning scented with hemp
saltily with bundle of clover.

Here my beloved hides for the night
safe from the boys in the kneading trough.
But up the ladder to the attic
carrying a measure of rye on his back
like a good husband

Here you are invited to the wedding.

(Leonard Kress)

SONG OF THE FIELD

I walked to the distant road
and I cast away my family home,
one hand grasping half the world,
so I can no longer return to the field.

But let the grizzled aspen grab me
somewhere by the boundary line
where gray hares haunch
and a wild blizzard rages.

110

Years and wars passed by me.
Rulers and schools passed by me.
My restless verse now slumps
on the castle's tables.

But let the grizzled aspen grab me
somewhere by the boundary line
where gray hares haunch
and a wild blizzard rages.

(Leonard Kress)

TADEUSZ RÓŻEWICZ (b. 1921)

Born in Radomsko, Różewicz saw service during World War II with the underground Home Army. Since then he has made a name as the single most influential Polish poet of the postwar period. His distinctive style of "naked poetry" has been followed by hosts of younger writers, the best among them Zbigniew Herbert. With Sławomir Mrożek Różewicz shares the distinction of being the most important Polish playwright of the postwar years. In the forties and fifties his major themes in poetry centered around the question of the possibility of art's existence after the horrors of war (as in *Niepokój* [Anxiety, 1947]); during the sixties he preferred to concentrate instead on problems of society in Poland and in the West (as in *Twarz* [Face, 1964]). In the seventies he wrote relatively little poetry. Różewicz, who lives in Wrocław, has been widely translated. Two excellent samplings of his work that appeared in the United States are *"The Survivor" and Other Poems*, translated by Magnus J. Kryński and Robert A. Maguire, published in the Lockert Library of Poetry in Translation series (Princeton University Press, 1976) and, more recently, the selection *Unease*, translated by Victor Contoski (New Rivers Press, 1980). At Mr. Różewicz's own suggestion the translations reprinted here are from this latter volume.

MOMENT

Poplars like grapes
on the old silver
of the cloudy sky

111

maybe here I'll find
what I really need

here's a girl
who went by
and with the most beautiful gesture
at this moment
on earth or in heaven
pushed back her hair
flowing with light
on her proud shoulders

mountains smoke
in the ash of heaven

I'll stay here
that's right
I don't need anything

Shadows on the walls
of small silent houses
and a bright shadow
on the balcony

the heart beats again
that died in that place
strange as a star
Verona

(Victor Contoski)

FORMS

Those forms once so well set
obedient always open to the reception
of dead poetic material
terrified by fire and the smell of blood
broke and dispersed

they turn on their creator
rend him asunder and drag
him down interminable streets

112

where long since
marched all orchestras
schools and processions

still-pulsing meat
brim full of blood
feeds
these perfect forms

so tightly converging over the spoils
not even silence
escapes

(Victor Contoski)

ROOTS

Photosensitive
aesthetes
with one eye
speaking of Van Gogh
paint suns
brush the banal branch
of the blossoming almond

I see him at night

I see him
in Borinage
underground
fire
devours people
with eyes
with beating heart
with tongue
surging through
the walled shaft

heaven is high
and rises higher
the eyeless mole
Van Gogh
touches light

113

looking at sunflowers
I think of roots
buried in the earth
pushing toward the sun
not knowing
light
crown

when a stranger greeted me
in the middle of the night

I knew him

(Victor Contoski)

TALENT

I sat by the wall
with eyes closed
face turned toward the sun
hands closed in a fist

in idyllic childhood
angelic diabolic childhood
I gave pennies
bits of bread
to beggars
they showed me stumps
split from the trunk
empty sleeves
open maws
scabs

finger by finger you pry open
two empty palms
turn the guts inside out
unwind the eyelids nothing

114

when my teeth were pried open
under the tongue was found
a black penny
alms from
the sun god

(Victor Contoski)

I BUILD

I walk on glass
on a mirror
that breaks

I walk on the skull
of Yorick
I walk on this crumbling
world

and build a house
a castle on the ice
everything in it
is prepared for the siege

only I am surprised
weaponless
outside the walls

(Victor Contoski)

REMEMBRANCE FROM A DREAM IN 1963

I dreamed
that Leo Tolstoy

lay in bed
huge as the sun
in his mane
of rumpled fur

a lion

I saw
his head
his face of surging yellow brass
where unbroken light
flowed

suddenly he went out
went dark
and the skin of his hands and face
was rough
broken
like the bark of an oak

I asked him
"what should be done"

"nothing"
he answered

by all his features
his cracks
light flowed toward me

a gigantic radiant smile
burst into flame

(Victor Contoski)

TYMOTEUSZ KARPOWICZ (b. 1921)

A writer active in several genres, as well as in literary criticism, Karpowicz took his doctorate in Polish philology at Wrocław University, where he also taught. Since 1974 he has been on the faculty of the University of Illinois at Chicago Circle. He began his career as a poet with his collection *Żywe wymiary* (Living Dimensions, 1948); there followed *Kamienna muzyka* (Stone Music, 1958), *Znak równania* (Equation Sign, 1960), *W imie znaczenia* (In the Name of Meaning, 1962), *Trudny las* (Difficult Forest, 1964), and *Odwrócone światło* (Reversed Light, 1972). Dr. Karpowicz is also a dramatist of stature, having written works for the stage as well as for the media; his work stands in the best absurdist tradition. His *Dramaty zebrane* (Collected Plays) appeared in 1975.

ON OPENING AN OLD HOUSE

the blind landscape gropes its way
to my one time house left forlorn
between an old birch which gave birth
to my visible world and a young stream
drinking from the hand
the dew of all daybreaks

what does it want from the boulder
which bars the door where stepfather
to this day is rolling a cigarette
although he died long ago

why does it push through mother's knees
who just at the doorstep
cooks sorrel with tears
thinner than us mother whom I once
buried among sorrel

I watch it press the door handle
purple from strain because it jammed
across the blindness
its face is final and it must come in
through mother and stepfather
and from behind the horizon lights
god rises unrelenting
in pressing forward on the dump
of brittleness of collected life
also finds gropingly the same
handle they both swell under the door
that has long stood ajar

what do they want or leave there
to take the abandoned by force

(Ewa Zak)

FROM THE SKEIN OF THE STRAIGHT LINE

how charmingly permanence strolls
it trots a few centimeters

117

near a crooked path then takes a turn
at a distance proper enough
from the road's bend and opens an umbrella at
three radii from the sun
tucks up the shadow under the frill
which skims behind it before the earth
drinks lemonade outside its lips
and smiles to a boy askew

philosopher on the track of the constant
crumbles bursts the binder of
the backbone from the amplitudes
of the rut into which he crept

luckily the broom of the ragman
sweeping up the integrated
utterly forgot itself among the leeways
of thought scribing a curve straight on

(Ewa Zak)

FROM BOSCH'S HILL

sow charlock around the ears
hearing the ruthlessly sweet voice
telling you about a sugar mountain
that swells in the liver of generations
for the glucoidal hallelujah of tomorrow

wire in the eyes before which stands
in pinkish flounces folded back
neatly fresh from pro-governmental tin
or demobbed pacts as far as the east
dawn with the passport from strangler's hand

and wrap up the tongue in fire
if it slides on the soap of statistics
not as a leister between hearts
but in the clitoris of the paid epilogue
held out by a water carrier
let there be ashes in place of lies
as the air for the enemy was black where
savonarola burned his hand

118

and the touch that withers on surfaces
of the former wolf's den now our own
on foreign drums braced tight
better strangle and on a touchless day
maybe a wailing wall will arise
which you will only pass at the sound of a die
the world's surfaces will follow you too

and the nose running deep into hogs after sausage
try to shove close to god's anus
if the smell in bosch's color will
sweep over you and the cry of mad marguerite
but your very own will pierce through charlock seeds
and the wire of gaze the knot of fire the noose of breath
blood will thrum the larynx will whirl in a void
it means you could still be thoroughly uncorked

(Ewa Zak)

artistic copy

*A MOMENT IN A SIDE AISLE
/with stained glass window/

stillness welling after speech
up the mineral icon of the air
sheds from the split lip
a pensive drop of blood

eternal is the sense of silence
of the stone prophet
his concrete tongue
now undermines all
ascending to heaven organs
on the viviparous wall
of the collective sin
bulges the vein of concentration

it is a votive
quotation from the fish
which haunted
by the satan of color

*Note: In this poem and in the two following it, successive sections justified at the right- and
the left-hand margins, respectively, should be understood as juxtaposed margin-to-margin, in
mirror-image fashion.—The Editor.

119

                              abandons its species
                              and lacking angels
                           with the chalice of joy
                              puts forth its gills
                          under ceaseless stanching
                                     of god

                                     amen

original

CLOTTING STOP

logical calque

DETERMINATIVE

the effective half of the bell on which
hangs the other half redundant belfry
to see this suspension the loveliest tolling
is when without a trace of sound one knows
what not to hear today to let resound all

(Ewa Zak)

                                   artistic copy

               INTRODUCTION TO A DESERT

                             in this desert
                       what riches of absence
                        having nothing to add
                                 to its cube
                     alienated from every picture
                     fata morgana with a languishing
                       lack of caravan in fact
                     leaves from the world's beginning
                      close behind the future track
                     the unborn flute of a bedouin

                                        PS
                    l'exactitude n'est pas la vérité
                    paints matisse on the cuff
              of the weather the needless wonder
      of god and the eye to be taken out by color meets
              its own entrance but passes it sideways

original

HURRICANE OF ABSENCE

logical calque

EMPTY TERM

milking a goat in seclusion
he has no view he borrows
from himself the edges of the pail
for the field and in his focus
passes the ocular habit
that one must see and be seen

logical calque

NOMINATING OBJECT

lack of basis for a place
piles it up from fickle habits
of carrying forth only
to have something to carry

(Ewa Zak)

REVERSED LIGHT

when my punt will reach the sun
i shall wait a while will not get off the shore
i shall look from here at all sets
of light at its names and synonyms
like unending scud uphill sweat
instant spurs in the eye a frantic
horseless rider lashing a branch
of precocious lilac stuck in another's horse the cheek
ploughed with pity the hauberk of order
in my restive arm the icy
seal of death on the snows of glory
better left untold like
a gnawed tunnel of a brotherly throat
where rats built nests out of manifestoes
my own hand grubbing in another's brain
in search of the formula for open life
in the hooded snare of death the scaffold on the altar
god in a battle dress with the defense a long
chopped off hand of memory on the sign

all that was indeed my reversed light
lit up when dying down and died down lighting up
but the nation's cluster on the heart kept skimming up the peak
of blood behind which i could never see the sun

original

NIBELUNGS IN A PIGSTY AS PULSARS

logical calque

INDIVIDUAL VARIABLE

symbol p as pasture but in such wind
that the predicate can barely hold
the sentence in hand for an hour in such
blasting into syntax to ask for the whereabouts
of the hat on your head one should not have
indeed even the full stop at the beginning of the phrase

122

logical calque

BOUND VARIABLE

often in barbed wire and in the middle of the globe
dragged by the knee length hair she was is
and will be my wounded homeland because from her i have
shoes for the daily roll call when i lace them up
i kink blood which does not want to ooze without me

(Ewa Zak)

MIRON BIAŁOSZEWSKI (b. 1922)

One of the most interesting contemporary Polish poets, in that he has
reexamined language and built a poetry out of an imagery of stark
everyday realism. His poetry of pipes and scaffoldings—of which some
examples are offered here—is a hallmark of his method. Among his
important volumes are *Mylne wzruszenia* (Erroneous Emotions, 1961),
Rachunek zachciankowy (Calculus of Whims, 1959), and *Było i było* (Were
and Were, 1965). Mr. Białoszewski is also a playwright and the author of
prose works; his important *Pamiętnik z Powstania Warszawskiego* (1970)
was published by Ardis in English translation as *A Memoir of the Warsaw
Uprising* in 1977.

From THEY'LL PAINT US

They clanged and clanged

Finally raised it.
They'll renovate us
on tubes.

The power of the ant-hill house

Scaffoldings higher and
higher.
Cosmos in the cage!

123

Scaffoldings
astonish

I anticipated—windows and cages.
I come back from my old house, open the door,
and here they are—cages, one cage
in fog
—beautiful!
—good!
early morning, can't see a thing in fog,
bits of two houses, can only hear the rest,
warm, submersion, suspension,
time tethered, protection,
boards from above guards from heaven,
in this cage as in a sheath,
as in medieval times,
you'd stand more boldly
& cry
"people, I'm a sinner!"

In a cage, on a cage

In scaffoldings little pigeons.
They lose feathers.
Feathers don't fall down.
Country women keep hanging out
dish-cloths, towels,
so what are they in cages?
Still the beginning of the day.
Craaash. Down it goes.
I'm coming back from where the trash-cans are.
A pale, pale-eyed is coming in
from the stairs side
—got any old papers?
I'm looking at him
as at wonderdawn
and only then do I say
—no.

Oh, it's literally tubes,
'cause I hear: tubes & tubes

when I still could see
they didn't have tube scaffoldings,
scaffoldings mean wood
—says Jadwiga blind for years.
Today we're going down the street
—there, here they are, touch 'em
—oh yeah, literally tubes,
and I took it metaphorically
—well, that's too
—yeah, but it's got legs

of tubes.

Warm,
night is fogging

from the moon
pours
honey
on tube
scaffoldings
we'll pull ourselves up
by the hair
till high
squeak
we'll come off
unglue ourselves
somehow our heavenly body
still lives, tempts,
don't slaughter it for the arrival.

(Tadeusz Slawek)

WISŁAWA SZYMBORSKA (b. 1923)

Szymborska was born in Kornik, near Poznań, but since age eight has lived
in Cracow, where she studied at the Jagiellonian University. She published
her first poem in 1945, has worked slowly and painstakingly since, and in
her introspective and yet cutting voice, has been one of the best-liked and
most-read poets of the postwar period. Her concise and understated style
are especially notable in her later collections, *Sól* (Salt, 1962) and *Sto*

pociech (A Barrel of Laughs, 1967). Her poetry has been popular abroad; substantial selections have appeared in both Germanies, and in the United States *Sounds, Feelings, Thoughts: Seventy Poems by Wisława Szymborska,* in translations by Magnus J. Kryński and Robert A. Maguire, has just appeared in the Lockert Library of Poetry in Translation series (Princeton University Press, 1981).

PREPARING A CURRICULUM VITAE

What is required?
Fill out the application
and attach a curriculum vitae.

Regardless of life's length
its outline should be short.

Mandatory are conciseness and selectivity.
Substitution of addresses for landscapes
and firm dates for shaky memories.

Of all the loves only the marital will do,
and of all the children only those actually born.

More important than whom you know is who knows you.
Mention travel only if abroad.
Membership in what but without the whatfors.
Honors but without the wherefores.

Write as though you have never talked with yourself
and have always steered clear of yourself.

Say nothing about your dogs, cats and birds,
your precious keepsakes, friends and dreams.

Rather the price than the value, the title than the contents.
Rather the size of the shoes than where he is going,
the person they take you for.

Attach a snapshot too, with one ear exposed.
What counts is its shape, not what it hears.

And what does it hear?
The clatter of machines turning paper into pulp.

(Magnus J. Kryński and Robert A. Maguire)

VIEW WITH A GRAIN OF SAND

We call it a grain of sand.
But it calls itself neither grain nor sand.
It gets along without a name that's general, particular,
ephemeral, permanent, erroneous or proper.

It has no need of our glance, our touch.
It does not feel itself perceived and touched.
The fact that it fell on the window-ledge
is merely our, not its experience.
That's the same for it as falling on anything
without the certainty that it's already fallen
or is falling still.

From the window there's a fine view of the lake,
but this view is unable to view itself.
Colorless and shapeless, soundless, odorless
and painless it exists in this world.

Bottomless to itself is the lake's bottom and shoreless its shores.
Neither wet nor dry its water to itself.
Neither single nor several feel themselves its waves
that murmur deaf to their own murmuring
around stones neither small nor large

And all this is under a sky by nature skyless,
in which the sun sets without setting at all
and hides without hiding behind an involuntary cloud.
The wind tugs at it for no other reason
than that it blows.

A second passes, another second, a third second.
But these are only our three seconds.

Time flies like a messenger with urgent news.
But that's just our simile.

An invented character talked into its haste,
and the news inhuman.

(Magnus J. Kryński and Robert A. Maguire)

TORTURES

Nothing has changed.
The body is susceptible to pain,
it must eat and breathe air and sleep,
it has thin skin and blood right underneath,
an adequate stock of teeth and nails,
its bones are breakable, its joints are stretchable.
In tortures all this is taken into account.

Nothing has changed.
The body shudders as it shuddered
before the founding of Rome and after,
in the twentieth century before and after Christ.
Tortures are as they were, it's just the earth that's grown smaller,
and whatever happens seems right on the other side of the wall.

Nothing has changed.
It's just that there are more people,
besides the old offenses new ones have appeared,
real, imaginary, temporary, and none,
but the howl with which the body responds to them,
was, is and ever will be a howl of innocence
according to the time-honored scale and tonality.

Nothing has changed.
Maybe just the manners, ceremonies, dances.
Yet the movement of the hands in protecting the head is the same.
The body writhes, jerks and tries to pull away,
its legs give out, it falls, the knees fly up,
it turns blue, swells, salivates and bleeds.

Nothing has changed.
Except for the course of boundaries,
the line of forests, coasts, deserts and glaciers.
Amid these landscapes traipses the soul,
disappears, comes back, draws nearer, moves away,

alien to itself, elusive,
at times certain, at others uncertain of its own existence,
while the body is and is and is
and has no place of its own.

(Magnus J. Kryński and Robert A. Maguire)

THE JOY OF WRITING

Where is the written doe running, through the written forest?
Will she drink from the written water,
that reflects her mouth like carbon paper?
Why is she lifting her head? Does she hear anything?
Poised on four slender legs borrowed from the truth
she flicks her ears under my fingers.
Silence—this word also rustles on the paper
and parts branches
evoked by the word "forest."

Above the white page letters lie waiting
to jump—they may be badly arranged,
sentences bringing to bay
from which there is no escape.

In a drop of ink there are many hunters
with squinted eyes
ready to rush down the steep pen,
to surround the doe with guns levelled.

They forget this is not life.
Black on white, other rules govern here.
The wink of an eye will last as long as I wish,
it may be divided into small eternities
full of buckshot suspended in flight.
Nothing will happen here, if I insist, forever.
Against my will no leaf will fall
nor grass blade bend under the dot of a hoof.

So is there such a world
I rule as absolute fate?
Time, which I bind with chains of letters?

Existence, at my command, never ending?

The joy of writing.
The ability to preserve.
Revenge of a mortal hand.

(John and Bogdana Carpenter)

INTERVIEW WITH A CHILD

Maestro has been with us since not long ago.
That is why he lurks in all the corners.
He covers his face with his hands and peeps through a chink.
Stands with his forehead to the wall, then turns back suddenly.

Maestro refuses with distaste an absurd idea
that a table not looked at must constantly be a table,
that a chair behind one's back remains in its limits
and does not even try to take the opportunity.

It is difficult to catch the world being different, that's true.
An apple-tree comes back to the window right before the blinking
of our eye.
Sparrow in rainbow colors always manage to turn grey in time.
An ear of the pitcher will catch any whisper.
A night table wears the passivity of the day table.
A drawer tries to convince Maestro
that it contains only what have been previously put there.
Even in a book of fairy-tales suddenly opened
a princess in the picture will always manage to sit down.

They feel I'm a stranger—sighs Maestro—
they don't want a newcomer to join their games.

Am I to believe that everything that is
exists only in one manner,
in a situation that is horrid since without a way out from itself?
without a break and change? Confined in its humble limits?
A fly in a trap of a fly? A mouse in a trap of a mouse?
A dog never let loose from a hidden chain?

Fire that can never afford anything else
but to burn again Maestro's trustful finger?
Is this a proper, ultimate world?
Wealth scattered not to be picked up,
useless luxury and forbidden chance?

No—shouts Maestro and kicks with all the legs
he has at his disposal—in such a great despair
that even the six legs of a lady-bug wouldn't be enough.

(Tadeusz Sławek)

WORDS

—La Pologne? La Pologne? It's awfully cold there, isn't it?
—She asked me and sighed with relief. There are so many countries nowadays, you know, that climate is the safest subject of the conversation.
—Madam—I want to answer her—poets of my country write in mittens. I wouldn't say they never take them off; they do if the moon swelters. They sing simple ways of the shepherds of seals in stanzas made of resounding shouts, as that's all that can be heard through the roaring storm. Classics carve with the icicle of ink on the beaten-down snowdrifts. Others shed tears of snowflakes over their fate, decadents. Those who want to drown themselves must have an axe to cut an air-hole in a frozen lake. That's the way it is my lady, my dear lady.

This is what I want to tell her. But I've forgotten what seal in French is. I'm not sure of icicle and air-hole either.
—La Pologne? La Pologne? It's awfully cold there, isn't it?
—Pas du tout—I answer and my voice is cold as ice.

(Tadeusz Sławek)

BORN

So this is his mother.
This small woman.
A grey-eyed originator.

A boat in which he sailed to the shore
many years ago.

It is from her that he was struggling his way
to the world,
to non-eternity.

The mother of the man
with whom I jump through fire.

So it is her, the only one
who did not choose him
as complete and ready made.

She caught him herself
in the skin that I know
and fastened to the bones
hidden from me.

She herself descried
his grey eyes
with which he looked at me.

So it is her, his alpha.
Why did he show her to me?

Born.
So even he was born.
Born like everybody.
Like me who will die.

A son of the true woman.
A newcomer from the depth of the body.
A traveller towards omega.

Exposed to his own absence
which comes from everywhere
at any time.

And his head
is a head against the wall
which yields only till the time comes.

And his movements
are abrogations
of the common fate.

I've understood
he has already covered half of his way.

But he did not say it to me,
he didn't.

—This is my mother—
he told me only that much.

(Tadeusz Sławek)

ZBIGNIEW HERBERT (b. 1924)

Zbigniew Herbert was born in Lwów in eastern Poland (now the Soviet Union) and studied philosophy at Warsaw. Since his first book of poems, *Struna światła* (A Chord of Light, 1956), he has published four additional collections, the most important of them being *Hermes, pies i gwiazda* (Hermes, Dog, and Star, 1957), *Studium przedmiotu* (A Study of the Object, 1961), and *Pan Cogito* (Mr. Cogito, 1974). Herbert is also the author of plays and of essays, primarily about culture and art. In 1970-71 he taught at the University of California, Los Angeles, and has also resided in West Berlin. He is currently living in Warsaw. His works have been translated into most of the principal Western languages.

THE RETURN OF THE PROCONSUL

I've decided to return to the emperor's court
once more I shall see if it's possible to live there
I could stay here in this remote province
under the full sweet leaves of the sycamore
and the gentle rule of sickly nepotists

when I return I don't intend to commend myself
I shall applaud in measured portions
smile in ounces frown discreetly
for that they will not give me a golden chain
this iron one will suffice

I've decided to return tomorrow or the day after
I cannot live among vineyards nothing here is mine
trees have no roots houses no foundations the rain is
    glassy flowers smell of wax
a dry cloud rattles aginst the empty sky
so I shall return tomorrow or the day after in any case I shall
    return

I must come to terms with my face again
with my lower lip so it knows how to curb its scorn
with my eyes so they remain ideally empty
and with that miserable chin the hare of my face
which trembles when the chief of guards walks in

of one thing I am sure I will not drink wine with him
when he brings his goblet nearer I will lower my eyes
and pretend I'm picking bits of food from between my teeth
besides the emperor likes courage of convictions
to a certain extent to a certain reasonable extent
he is after all a man like everyone else

and already tired by all those tricks with poison
he cannot drink his fill incessant chess
this left cup is for Drusus from the right one pretend to sip
then drink only water never lose sight of Tacitus
go out into the garden and come back when they've taken
    away the corpse

I've decided to return to the emperor's court
yes I hope that things will work out somehow

(Czesław Miłosz)

ELEGY OF FORTINBRAS
    for C.M.

Now that we're alone we can talk prince man to man
though you lie on the stairs and see no more than a dead ant
nothing but black sun with broken rays
I could never think of your hands without smiling
and now that they lie on the stone like fallen nests
they are as defenseless as before The end is exactly this

The hands lie apart The sword lies apart The head apart
and the knight's feet in soft slippers

You will have a soldier's funeral without having been a soldier
the only ritual I am acquainted with a little
There will be no candles no singing only cannon-fuses and
    bursts
crepe dragged on the pavement helmets boots artillery horses
    drums drums I know nothing exquisite
those will be my manoeuvres before I start to rule
one has to take the city by the neck and shake it a bit

Anyhow you had to perish Hamlet you were not for life
you believed in crystal notions not in human clay
always twitching as if asleep you hunted chimeras
wolfishly you crunched the air only to vomit
you knew no human thing you did not know even how to
    breathe

Now you have peace Hamlet you accomplished what you
    had to
and you have peace The rest is not silence but belongs to me
you chose the easier part an elegant thrust
but what is heroic death compared with eternal watching
with a cold apple in one's hand on a narrow chair
with a view of the ant-hill and the clock's dial

Adieu prince I have tasks a sewer project
and a decree on prostitutes and beggars
I must also elaborate a better system of prisons
since as you justly said Denmark is a prison
I go to my affairs This night is born
a star named Hamlet We shall never meet
what I shall leave will not be worth a tragedy

It is not for us to greet each other or bid farewell we live on
    archipelagos
and that water these words what can they do what can they
    do prince

(Czesław Miłosz)

135

NAKED TOWN

On the plain that town flat like an iron sheet
with mutilated hand of its cathedral a pointing claw
with pavements the color of intestines houses stripped of
    their skin
the town beneath a yellow wave of sun
a chalky wave of moon

o town what a town tell me what's the name of that town
under what star on what road

about people: they work at the slaughter-house in an
    immense building
of raw concrete blocks around them the odor of blood
and the penitential psalm of animals Are there poets there
    (silent poets)
there are troops a big rattle of barracks on the outskirts
on Sunday beyond the bridge in prickly bushes on cold sand
on rusty grass girls receive soldiers
there are as well some places dedicated to dreams The cinema
with a white wall on which splash the shadows of the absent
little halls where alcohol is poured into glass thin and thick
there are also dogs at last hungry dogs that howl
and in that fashion indicate the borders of the town Amen

so you still ask what's the name of that town
which deserves biting anger where is that town
on the cords of what winds beneath what column of air
and who lives there people with the same skin as ours
or people with our faces or

(Czesław Miłosz)

THE FATHERS OF A STAR

Clocks were running as usual so they waited only
for the avalanche effect and whether it would follow
the curve traced on a sheet of ether
they were calm and certain on the tower of their calculations
amid gentle volcanoes under the guard of lead
they were covered by glass and silence and a sky without

secrets
clocks were running as usual so the explosion came

with their hats pulled tightly over their brows they walked
    away
smaller than their clothes the fathers of a star
they thought about a kite from childhood the tense string
    trembled in their hands
and now everything was separated from them
clocks worked for them they were left only
like an heirloom from father on old silver pulse

in the evening in a house near a forest without animals or
    ferns
with a concrete path and an electric owl
they will read the tale of Daedalus to their children
the Greek was right he didn't want the moon or the stars
he was only a bird he remained in the order of nature
and the things he created followd him like animals
like a cloak he wore on his shoulders his wings and his fate

(Czesław Miłosz)

MR. COGITO CONSIDERS THE DIFFERENCE
BETWEEN THE HUMAN VOICE
AND THE VOICE OF NATURE

The oration of the worlds is untiring

I can repeat all of it from the beginning
with a pen inherited from a goose and Homer
with a diminished spear
stand in front of the elements

I can repeat all of it from the beginning
the hand will lose to the mountain
the throat is weaker than a spring
I will not outshout the sand
not with saliva tie a metaphor
the eye with a star
and with the ear next to a stone
I won't bring out stillness
from the grainy silence

137

and yet I gathered so many words in one line—longer than all the lines of my palm and therefore longer than fate in a line aiming beyond in a line blossoming in a luminous line in a line which is to save me in the column of my life—straight as courage a line strong as love—but it was hardly a miniature of the horizon

and the thunderbolts of flowers continue to roll on the oration
                                                           of grass the oration of clouds
choruses of trees mutter rock blazes quietly
the ocean extinguishes the sunset the day swallows the night
                                                           and on the pass of the winds

new light rises

                              and morning mist lifts the shield of islands

(John and Bogdana Carpenter)

SENSE OF IDENTITY

If he had a sense of identity it was probably with a stone
with sandstone not too crumbly light light-gray
which has a thousand eyes of flint
(a senseless comparison the stone sees with its skin)
if he had a feeling of profound union it was exactly with a stone

it wasn't at all the idea of invariability the stone
was changeable lazy in the sunshine brightened like the moon
at the approach of a storm it became dark slate like a cloud
then greedily drank the rain and this wrestling with water
sweet annihilation the struggle of forces clash of elements
the loss of one's own nature drunken stability
were both beautiful and humiliating

so at last it would become sober in the air dried by thunder
embarrassing sweat the passing mist of erotic fervors

(John and Bogdana Carpenter)

WIKTOR WOROSZYLSKI (b. 1927)

Woroszylski broke into print right after World War II as one of the *pryszczaci* ("pimpled ones"), a contemptuous name bestowed by their elders on those members of the rising generation of writers who were fanatically devoted to the theory and practice of socialist realism. The very title of a collection of his verse, *Śmierci nie ma* (There Is No Death), illustrates his cast of mind at the time. From 1950 to 1952 Woroszylski was one of the editors of *Nowa kultura*. He then went to Moscow, where he graduated from the Gorky Institute of Literature. His selected poetry, *Wybór wierszy* (from which the poems here offered are taken), appeared in 1974. Since then he has been writing a regular column for the Catholic monthly *Wieź* (The Bond).

THE EARLY CHAPLIN

We know the early Chaplin but before him
was an earlier Chaplin Those films
are already forgotten Chaplin
had the face of a brute he bared his teeth
like a wildcat ready to spring and had no
scruples he seduced women hired gangsters
to rub out his rivals rejected he sought revenge
by all means fair and foul when the lovely Mabel
Normand was in an auto race this blackguard
wetted down the pavement switched around the signs sending
the car to the precipice planted dynamite despite
everything Mabel survived and won he almost
burst with fury the audience
was bursting with laughter and of course showed
him no sympathy at all He did not yet have
his dress suit cane mustache He had
a monocle and goatee His name was not
Charlie but Chas This went on
for some time But in time
his face figure and inner self began to change This
happens not only to actors but
to apostles and mere mortals More than one
person has lived through this in his youth wondering
and suffering when his skin
hardened in a grimace which
seemed to be an exact replica

of the inner self when it crumbled and again froze
in another mould No one knows
how all this happens This is no
run-of-the mill decision Maybe inspiration Thus Chaplin
after making thirty-five films went ahead and
changed from Chas to Charlie In the last
of these films they both were there Charlie
dreamed of being Chas From sleep
the loiterer was wakened by the policeman's stick He got up
from the park bench and hobbled
into the distance with an embarrassed smile There was no
anger in him only hope No craving
for success but a desire
to defend human dignity From the pursuer
he became the pursued And this was now
the early Chaplin whom
we came to love without ceasing to laugh

(Magnus J. Kryński and Robert A. Maguire)

FRANZ KAFKA (1)

When Franz Kafka was writing his stories they were not
a mirror of reality He was born and lived
in the mildest of tyrannies and after it fell in the most
decent of the bourgeois democracies of Europe In a world
so unreceptive to Apocalypse that when swallowed up
by it could not cease wondering In the city where
to this day one can find traces of everyone who lived there
before the Apocalypse Also
of Franz Kafka Here is
the house where he was born Here is where the servant girl
told him scary stories on the way to school Here
is where he worked in an insurance company and did not dare
take two days off to leave the city
and meet his mistress Here is
the Kafka family tomb All the houses
are still standing but from some
the plaster is falling The tenants
also point out the old well he described
in one of his first tales and the interior of a church
from a novel But

140

if one were to seek in what he wrote a picture
of the world in which he lived that would be
unjust and that honorable world would have a perfect right
to feel itself slighted because it was not
like that It was
more serene and simple Fortunately
they started reading Kafka much later when the world's body
began to creep into each of his invented words making up
the time lag and moaning with pain.

(Magnus J. Kryński and Robert A. Maguire)

FRANZ KAFKA (II)

Franz Kafka always wished
to be trusting and inspire trust Among
the books he read most avidly were
the stories of Božena Němcová Often
he returned to them absorbing
the warmth that they gave off and wishing
to become like the good old lady who distributed
bread to people Once
he read a work of hers entitled *In the Castle
and the Village* and this made on him
such an impression that he could not
resist the desire to imitate He wrote
long and kept hoping he'd succeed
in creating an equally sunny work with an equally
happy ending but
he did not succeed and suffered greatly and wanted
to burn what he had written.

(Magnus J. Kryński and Robert A. Maguire)

In the Castle and the Village is the title of a semi-fictional prose work by Božená Němcová. The Czech is V zámku a v podzámcí.

TADEUSZ NOWAK (b. 1930)

Born of peasant stock, Nowak has retained a lifelong interest in the themes and images of his background. In his volumes of poetry, which have been appearing since 1953, he is preoccupied by village life, rural childhood, folklore, myth and ritual, ancient modes of behavior, and the freshness of the perceptions of children. He has also written novels, among them *Obcoplemienna ballada* (Ballad of an Alien Tribe, 1963). He has been called by Miłosz a "peasant surrealist."

PSALM OF THE FIELD

I write a psalm in the field Corn sheds blossoms
and shorn grass scents the pre-dawn air
Dew will trickle from the knife drop after drop
The crucified body held up by a spear

Only the dog staring fixedly at the sky
from his hut hears the other world and barks
though through the heavens viaticum thin
nothing but a few bones of man can be seen

Only these feathers pouring forth from the hay
fluffing at the bank of the river
bear witness that the village prayed in front of someone
throughout the night for protecting weapons

Only the fool who contemplates the holy day
and sees the sacred Magdalen in the myrtle
and hugs the horse's mug the ancient way
whispers about my unimaginable poverty

I write a psalm in the field Hay withers
in the sun Standing in the river horses drink the water
Already now in the village no one can be tormented
and my psalm veils itself with lilies

(Leonard Kress)

PSALM TO CHEER UP

And his bleeding body beside the wound of morning
snuggles up to the drowsy hay
and the horse-grass and the dog-herbs
where a grassy god truncated by the scythe calls out

And the body does not know which hand is right
and the body does not know which hand is left
But on each side grass in the sun is drying
and from each blade God mowed-down is singing

And he hears the morning beside the wound
in the grass to whom life is prescribed
and to whom eternity malleable as honey
will satiate our clandestine hunger

And he sees the body beside the wound of morning
that will pick itself up from the hay like a grassy god
and bleeding run in the birches where in the white morning
he will fling himself on the breast of a young girl

(Leonard Kress)

PSALM AJAR

That's not how to love But to drink vodka
To slit the cock's throat before Rosaries
After daybreak three Marys enter the wound
to kiss the tin jaundiced body

And within him everything comes ajar for you
asleep and running up to the steps of consciousness
The body of Judas appears so near
a razor can be heard slashing the artery of water

(Leonard Kress)

143

PSALM OF LOVE

I stand by you   The horse heaves a sigh cross the river
Hay pours forth from a morning of haymaking
and instead of the galling grackles the orchard reigns in heaven
serving apples on the morning's spear

I stand by you   Cain beside Cain
now that Abel is dead and burned
on your neck and on my neck
a medallion in which his ashes are set

I stand by you blade by blade
between us the remains of our fate: feather of grass
earth sky: like the simplest doll
to be given back to the ancient ones to play

(Leonard Kress)

JERZY HARASYMOWICZ (b. 1933)

Jerzy Harasymowicz is a poet in whose work, as Miłosz writes, "surrealism, stifled in Poland during 1918-1939 by the rationalistic First Vanguard, took belated revenge" (*The History of Polish Literature,* p. 481). Whether we wish to call Harasymowicz's work surrealistic, primitive, anachronistic, or simply childlike, it is certainly filled with the wonders to which the title of his first work, *Cuda* (Wonders, 1956) is an appropriate tribute. Harasymowicz has been a widely-read and popular poet; his *Wybór wierszy* (Selected Poems, 1967) contains examples of his longer poems, in which, among other subjects, he celebrates the city of Cracow, where he lives. Poems from an earlier volume, *Genealogy of Instruments* (1959), have been set to music by Krzysztof Penderecki.

WOODS

Poison mushrooms
want to be picked
at all costs

Green undergrowth
cuts the hand
like a razor

Marshes try
to sell their carpets
to everybody

In a clearing
black with poison

Hemlock and hellebore
smile at you

The woods are quiet
and gentle

(Victor Contoski)

WAKING UP

First of all
the old icon clears its throat

Then
the stove rings

Then
the kettle crows

Then
the table is harnessed
the chairs kick up their heels

And coffee smokes
and a new day
opens its knife

(Victor Contoski)

From the Sequence UMBRELLAS

OLD AGE

A sad old
umbrella

looks
with blue
faded eyes

kneeling helpless
on a couch
full of books

it sleeps continually
now
folded under the ceiling
with its doggy muzzle

once
I visited it
still opened

a small pink umbrella
of Chagall
in its buttonhole

but
through the holes
in the lungs

the silver wires
of the ribs

shown already
much too clearly

(Victor Contoski)

RYSZARD KRYNICKI (b. 1943)

Krynicki graduated in Polish philology from the University of Poznań; his first appearance in print as a poet was in 1966 in the monthly journal *Nurt* (The Current), of which he was also an editor for several years during the sixties. Krynicki's poetry shows traces of influence by Rózewicz. However, he speaks with a voice of his own, and has already established himself as a leading poet of the younger generation. It may not hurt, however, to note that charity soliciting as portrayed in "By Entering" is a by no means unusual occurrence in the United States.

BY ENTERING

"By entering the grand lottery
of the Center for Children's Health
you honor the memory of the 2,000,000 children
who perished in battle and were brutally
murdered in World War II.
You bring help and relief
from suffering to thousands of children
crippled and afflicted with disease.
You fulfill your noble and honorable
civic duty.
You have a chance to win
many valuable prizes such as
the Fiat 125p and 126p car,
the C-330 tractor, television sets,
radios, refrigerators, sewing machines,
washers

and a host of other attractive products."

(Magnus J. Kryński and Robert A. Maguire)

The poet says in a note: "Posters like this actually did appear."

EWA LIPSKA (b. 1945)

The youngest of the Polish poets here represented, and certainly one of the most talented. Born in Cracow, where she continues to make her home, she pursued studies in the visual arts at the Cracow Academy of Fine Arts. Since 1969 she has been poetry editor at the Cracow publishing firm Wydawnictwo Literackie. Five collections of Lipska's poetry were published by Czytelnik of Warsaw between 1967 and 1978. A selection of these poems, under the title *Dom spokojnej mlodośći* (The Home of Tranquil Young Age) was published by Wydawnictwo Literackie in 1979, the year in wich she was also awarded the prestigious Robert R. Graves Award of the Polish P.E.N. Club.

WHEN OUR ENEMIES FALL ASLEEP

When our enemies fall asleep
we surprise them from the rear
open wide the gates of their skulls
lower the drawbridge
of frontal bone
along which enter truckfuls
of the fresh meat of conscience
the green parsley of new ideas
and the frosting on the cake of imagination.

In the morning
we sign with them
a pact of non-aggression.

(Magnus J. Kryński)

IF THERE IS A GOD

If there is a God
—I'll have dinner at his place.
Instead of a stop light, a red hawthorn.
An angel will come for me in a car.
Doves of chubby clouds
will flutter over the folding table.
From empty jugs we will drink
holy water and free will.

Even if God is nearsighted
he'll see eternity coming.
If God has a flair for languages
he can translate holy poems
for an anthology even holier
than the holiest first drop
from which a river sprang.

Later we will go cycling, God and I
over a cherry tree, over the landscape of paradise.
Earth's reeds stand in vases here.
Beasts of prey lie fallow.

At last God will get off his bike and say
it is he
who is God.

He will take out his binoculars. He will command me
to behold the earth. He will tell me
how things have come to such a pass,
how long he has plied his trade
and how infallibly he has failed with the world
launching tiny airplanes of ideas into the vacuum.
If God is a believer
he prays to himself for perpetual hope.

Oxen carry the sun upon their horns.
The folding table sways on its legs.
I will get medicine from God
and get well
soon after I die.

(Magnus J. Kryński)

WHAT HAPPENED TO PROFESSOR WHITE
ON LEAVING THE HAT STORE

Professor White
who switched the heads of two chimpanzees
bought himself a fashionable felt hat.
On leaving the store
the hat leaned down to his ear
and whispered:

Watch out, White,
the track of your brain
is the site of a track meet.
Your thought made a long jump
a jump far too long
for this mankind of ours.

(Magnus J. Kryński)

THAT MOMENT

That fluffy squirrel
suddenly has something human to tell me
that butterfly fluttered near me
not by mere coincidence.

A bird looks at me
with the eye of someone I know.

The thrush's call warns me
against killing time.

I know too well the silence
that falls from the owl's beak

The grimace of the tired lioness
behind bars that are all too human

The dog's freedom
that looks me in the eye with the trust
of entire nations

That wolf's howling. The suicide of the moth.
Someone dear to me must be calling for help

That moment
comes to me suddenly
and batters my heart
against cold air.

(Magnus J. Kryński)

EAST GERMAN

INTRODUCTION TO EAST GERMAN POETRY

It is an interesting thought that if Goethe and Schiller were alive today they would be East German writers. The role would probably become revolutionary Schiller better than it would conservative Goethe; but the backgrounds of East German writing are to be sought in literature both more and less recent than the youthful storm-and-stress dramas of either classical giant. Brecht's baroque sources include, as we know, John Gay, Villon, and the materials of the Thirty Years' War; much more recently, Gerhart Hauptmann, Frank Wedekind, and Carl Zuckmayer have stirred sensibilities and consciences with social dramas written, in some instances, before World War I. From the present vantage point of East German poetry we may feel that even Brecht's influence, in poetry as well as in drama, was clearly prophetic as well as instigative at a time when few were looking far enough ahead to see the Wall. Strange as the suggestion may seem, lyric poetry is perhaps the ideal genre to show now that East German literature is in some significant ways Western in origin and ongoing aspirations both. The nature poets Huchel and Bobrowski have correctly been identified as being of the visionary line that extends from Klopstock and Hölderlin to Trakl and to Rilke, and it would not be too meaningful to speak of the minimalism of Brecht and Kunze without relating it to corresponding tendencies in the late poetry of Celan. Yet these well-established facts still leave us within the confines of German literature. Part of the paradox of the inverse fortunate fall of East German man may well be that he has also been forced to pay some attention to the classics of Russian and East European poetry. Out of this alchemy, too, there may have emerged a peculiarly East German poetic consciousness, and at last a respectable academic subject which the professors here are fond of referring to, somewhat macaronically, as "DDR poetry."

There is no "East German" language, in a sense very different from that in which there is no "South Slav" (i.e., "Yugoslav") language; the contemporary German idiom is held in common between East and West. That elementary fact alone has, during the period of our interest, encouraged considerable literary traffic over the Wall. A third point to keep in mind (perhaps the least obvious of the three) is that the harassment and bureaucracy that we have heard about may not be all counterproductive: the need to put up resistance can also be a goad to achievement. In any case individualism, soil of the true poetry of private and shared experience, has remained fertile under the auspices of reluctant authority that banishes some of the best writers while bestowing the Fontane and Heinrich Mann Prizes on others (or, under changed circumstances, on the same individuals). Now that a literature has come of age, and a second generation of those born and raised under the system are raising their voices, it seems to be time to attempt another focus, a broader perspective. It does not take bureaucrats to tell a writer that an ideology of radical individualism is untenable. Roland Barthes, one of the most Western-oriented semioticians, has shared with us the awful secret that nothing that we write is original—there is no "creativity" as such. We forge, out of materials already shaped symbolically, segments of the metals of language that at best bear the imprint of our decisions. This leaves plenty of room for working in what we might call that "subtler collective," and for reasserting and freshly sharing, as East German poets surely have done, our common humanity.

The poets represented in this section do not, unfortunately, include all of the good names, and I feel that our inability to represent Volker Braun, Adolf Endler, Heinz Kahlau, and Karl Mickel is our loss. The roster does include just under one dozen of the best people even so. The work of three groups of writers is being stressed, with unequal degrees of emphasis: the classical nature poets (Arendt, Huchel, Bobrowski) and members of the younger generation (Czechowski, Kirsch, Bartsch) form, as it were, promontories, with the valley of the minimalist poet (Kunze) lying between. Nature poetry, the suggestive resonance, for example, of Huchel, is again becoming important, and especially Arendt's and Bobrowski's image systems can suggest to the novice a strange sense of place. The turn to the East is sensed in Bobrowski's Baltic images, in his sense of irretrievable loss, in life if not in vitality, under places presided over by water and by fire. The ravages of time and of history leave "the sky . . . / open, in the color of a child's hair. / Beautiful earth fatherland" ("The Deserted House"). How much pain and disappointment in that word "fatherland." Just so can another borderland spirit, Hanns Cibulka, utter: "Land, / I have half-forgotten / your language." His voice reverberates with Hölderlin's "we have all but / Lost, in foreign parts, language itself." In Arendt's hymnic language the very un-German presence of sea and beach, far north and east, is experienced near a "lost bay"; nothing can restore this sense of loss, no Bobrowskian sanctuary, not even Huchel's rare and comely affirmation: "Our homeland is beautiful / when a crane calls out over the green / brass circle of the pond / and gold is hoarded / in the blue cellar of October."

Traces of Brecht's influence remain with East German writing, and Reiner Kunze's minimalist poetry, with the volume turned down, with its intimacies and cruelties, well represents it (in this connection see especially Michael Hamburger's introduction to *East German Poetry*). Günter

Kunert's poems printed here, while physically more generous, also tend to be minimalist in spirit. They tend to deal in general truths rather than in the specifics of our days and feelings. In contrast the poems of Heinz Czechowski, Sarah Kirsch, and Kurt Bartsch offer a fresh voice and a healthier relationship with language, although these three poets were all born within one-half decade of Kunze. Kirsch and Bartsch share shy bravado, a devil-may-care overtone and yet a core of deep caring. "Drinking Vodka" displays Kirsch's surprising sensitivity to images and textures, before the dulling of the senses brings out perceptions of an insulted brain ("we'll walk on polished roads / that bend around the mountains"). Perhaps Bartsch's "Rita" expresses the human predicament most directly: "What should I do / I'm over forty, divorced, three kids on my back / The youngest is thirteen and goes with men already." There is little sense in blaming systems, neighbors, even ourselves: "We dug the city out with bare hands / Now the city stands. I'm a ruin." Oh yes, East German poetry does tell us what we hold in common—the power to survive and the threat of extinction are angels wrestling within each and every one of us. We understand this strange diction because, so far, the angel of survival has won the rounds.

ERICH ARENDT (b. 1903)

Erich Arendt published his first poems in Herwarth Walden's journal *Der Sturm* in 1926. He joined the Communist Party in the same year and was an early member of the League of Revolutionary Proletarian Writers (1928). In 1933 he emigrated to Switzerland, and from 1936 to 1939 he fought in the Spanish Civil War with the 27th Catalan Division "Carlos Marx." After a short period spent in France Arendt moved to Colombia, where he spent his exile years, and from here he returned to the German Democratic Republic in 1950. Arendt is a lyric poet of great versatility and power, as well as a fine translator (of Neruda, Asturias, Aleixandre, Alberti, Whitman, inter alia). He has also produced some impressive travel and art books.

LOST BAY

Of ashes the lifeless land.
Sand,
a sea-grey ashen land.
The cacti in the lightless light

155

carry
the heavy iron weight:
the paling wall of the sky,
wall
which consists of dead light.

Death haunts the shore, sere and dread:
thorns
stiffen to shapes that terrify,
A single blossom blossoms red,
dies
from the stabs of the cliffside thorn.
Thorns clog the breath, block the eye.
Wind
wanders past, forlorn.

Bones that are washed from the sea:
tree
upon tree, strange and primevally
huge, from the woods of the sun-flood.
Trees,
which stood, cast in hard wood,
millennia deep in the hot
glow,
wood, fit only to rot.

The waves move in, rolling and heavy.
Fear
is trembling through the moonbare bay,
and the shark's fin cuts past pale-grey.
Clouds still
throw their shadows in the sea.
For years no songbird's passed this way.
Sea
and its sky are empty.

Of ashes the lifeless land.
Sand,
a sea-gray ashen land.
The world's budding growth palls,
but
there on the hillside a negro. He hauls

the wooden plow by hand.
Weighs
every clod as it falls.

(David Scrase)

ODE I

Earthly bare,
How it blows you! dumb
from the deepest age of the world: and faceless,
a thinking, desolate,
from cliffs and moonless tides!
Before the hard wings of the light,
walling up impenetrable skies,
the world waves of stone: You
timelessness: rigid
horror! where no man ever
suffered his hour
nor ever looked up, hoping.

You who are born of flight, were not
those whom dawn awoke with
gentle and tapping knuckles,
remember, were not the eyeshining heavens
round you? Once as you stood
in the bow of the solitary drifting floe, of our night —
born earth, the swallow's wing,
a shadowless finger, came and swept
steeply across the sea which stretches out to
death, and touched you and
aroused in you, as later distance
in the traveller, the painful
blossom...

Hear her again, whom
the trader in dust at the crossroads,
the keeper of the freezing grave,
denies to you day by day;
hear her,
the lark as she soars aloft, childhood,
a singing, high,

from her hyacinth-colored minaret!
...and when there was only sky, and stars
played more consolingly, a praising
from shells and grass
the astonishing night...still
over the ready sleep of
the world the blissful buzz of insect wings:
Ripening light
of an inwardness!

Look earthwards, flying one, look:
O inextinguishable seal, kiss
from the lips of the reddening dawn
to the darkly circling sphere!
For which—there's always
a longing—it's a matter of going
home, beneath the icy sheen of the sky
where death's ecliptic
topples moons
and, time being ripe,
the solidifying heart of the sun;
for which,
this sphere, lost in the universe,
you owe your love, seeking the dead
secret of the birds,
the long since vanished trace.

Once above the clouds' silence
large as the sea everything
was flight. And the arc was
dark with the rushing.
By the sides of the last hill
dragging shadows halted: anger,
black winging blood. At night—
how near still, man, near your
endangered heart! haunting fingers
from cave and swamp: flapping ones!
Oh! over your face, o earth, naked
and formless the periosteal beings glided,
shadow
on shadow, till, you who breathe again,
it sank in the fossilized twilight: blood
coagulated fear.

It lay yet on the faces
of the cleft and splitting rocks
late-born and somber angel
in the unbroken wind:
Petrified mask of a cathedral.

Did the tear, touching dust and flower, not fall
as well? Think: all of the fruitless
longing, fields of dead pale behind
you, a wood which is felled
by all the dark.

Melancholy has lain from the
very first on the shady hills of our memory.
And before your eyes
the naked honest cliff: Never
under sinking heavens was
a dying well-accomplished. Only now and
then, in the creative death
of days, the hollow
wave threw a radiance, quivering
in one glance, as if born afar,
a smile; Rarity!... which
saw the shining cloud
drift off—but where?

Eyes of the dreamer
for ever closed, shall I then
lament in all the stone and
drifting sand of the hours?
White umbels drank deeply
of deadly bitter solitude on the blind bank,
as life does always
of transience. But, that which was
designed to foresee in infinite space,
a heart, didn't it perceive,
as if the sorrowful sea were closing,
the more delicate heavens'
appassionata?

Unmeltable heart
of feeling! For all
the dolphins might
sing from the bare ocean

into the dead light,
for One in the view of the world
the hour opened up:

lightly budding wind
you, at the eye of the waters:
breathing one!
Mountain of waves, flown over
by the first fish:
dream
of wings and fins.

(David Scrase)

PETER HUCHEL (1903-1981)

Peter Huchel, until his death in April of 1981 widely regarded as the premier East German poet, was born in Berlin-Lichterfelde. Huchel's poems began appearing in the twenties, and his first book of poems, *Gedichte,* came out in 1948 in East Germany, followed by a West German edition in 1949. Immediately following the war Huchel was director of East Berlin Radio, and between 1949 and 1962 he served as editor-in-chief of the influential literary journal *Sinn und Form,* which many critics have referred to as the most important literary periodical in the German-speaking countries. In 1971 Huchel settled in the West, and spent a year in Rome; subsequently he lived at Stauffen, near Freiburg im Breisgau, West Germany. Huchel's books have appeared in the West; among them are *Chausseen Chausseen* (Highways Highways, 1963), *Die Sternenreuse. Gedichte 1924-1947* (The Star Trap: Poems 1924-1947, 1967; a revised edition of the 1948 *Gedichte*), *Gezählte Tage* (Numbered Days, 1972), and, most recently, *Die neunte Stunde* (The Ninth Hour, 1979). Huchel's art is a dense and complex one, creating a texture of symbols and feelings out of nature motifs and language evoking human isolation.

THE SPINDLE

Sunken road,
the sycamore pinion rustling above,
And at the top of the slope
the gigantic spindle,

160

the hay of the peasants
piling up around a shaft.

Who hides alone
in rumbling steam,
in the cloak of the night?
Who turns the spindle
near the wind-blown grass?

A thatched roof
thrusts its wedge into the sky.
I see it spin,
ancient,
near the cooking fire,
the spindle gathering dust from the bran,
the long days
in stable and shed.

It hums, the spindle behind the forehead,
and winds
the fiber of plunging years.
Salt blows in the leaves of the funeral wreath.
Taxes and bribes
disperse in the rain.

On the hearth the boots warm themselves.
They bend and count
the ties of the military railway,
the snow-frosted milestones.

Smokeless flames flicker higher.
And outside, the night,
the raw liturgy of the winds,
branches breaking over the tombstones.

Spindle on the slope,
your fibers blow cold.
But I carry smoldering embers,
the promise of the dead,
through the sycamore gloom of the canyon.

(Rich Ives)

THE POPLARS

Time with your rusty scythe,
it was late when you traveled on
along the sunken road
and past the two poplars.
They swam
in the sky's thin water.
A white stone drowned.
Was it the moon, the eye of solitude?

Dusk on the graveside bushes.
It wound its cloth,
thickly woven with grass and mist,
around helmets and bones.
The first light, encrusted with snow,
cast glittering fragments into the reeds.
Silently the fisherman pushed
his boat into the river. The freezing
voice of the water complained,
corpse after corpse floating down and away.

But who buried them in the frosted clay,
in ashes and mud,
the old footprints of necessity?
Amid leveling war, fields glisten,
the swelling strength of a stem presses upward.
And where the plough turns
the stubble under,
the two poplars are still on the slope.
They tower into the light
like the earth's antennae.

Our homeland is beautiful
when a crane calls out over the green
brass circle of the pond
and gold is hoarded
in the blue cellar of October;
when corn and milk sleep in the storeroom
sparks fly
from the anvil of night.
The world's sooty forge
begins to fan its fire.
It hammers

the glowing horseshoe of dawn.
And ash falls
on the shadows of bats.

(Rich Ives)

THRACE

A flame tongues
here on the ground at night,
it whirls white leaves.
And at noon shatters
the sickle of light.
The rustling of sand
cleaves the heart.

Do not lift up the stone,
that warehouse of silence.
Beneath it
sleeps the centipede
of time.

Over the pass,
notched with horses' hooves,
blows a mane of snow.
With the smokeless shadows
of numerous fires
evening fills the canyon.

A knife
skins away the fog,
the battering-ram of the mountains.
Across the river
live the dead.
This speech
is their ferry.

(Rich Ives)

THE AMMONITE

For Axel Vieregg

Tired of the gods and of their fires,
I lived without laws
in the dip of the valley of Hinnon.
My old companions left me,
the balance of earth and sky,
only the ram, trailing its footrot limp
across the stars, remained loyal.
Under its horns of stone
that shone without smoke, I slept by night,
every day baked urns
that I shattered against the rock
in face of the setting sun.
In the cedars I did not see
the cats' twilight, the rising of birds,
the splendor of water
flowing over my arms
when in my bucket I mixed the clay.
The smell of death made me blind.

(Michael Hamburger)

ON THE WAY

Day felled
the roaming troop
of ice-covered leaves
with wires above the fire pit.

Beside her wagon,
sheltered by its cover,
the Gipsy woman,
her swaddled infant
asleep at her feet.
From the sheepskin she raises
a puppy to her breast;
suckling it,
she suckles the hungry wind in the snow.

Distant daughter
of the Asiatic goddess,
you've lost your flint sickle
on the edge of the hellish marshes.
By night you hear the barking
that follows the wheel tracks from camp to camp.

(Michael Hamburger)

EASTERN RIVER

Do not look for the stones
in water above the mud,
the boat is gone.
No longer with nets and baskets
the river is dotted.
The sun wick,
the marsh marigold flickered out in rain.

Only the willow still bears witness,
in its roots
the secrets of tramps lie hidden,
their paltry treasures,
a rusty fishhook,
a bottle full of sand,
a tin with no bottom,
in which to preserve
conversations long forgotten.

On the boughs,
empty nests of the penduline titmice,
shoes light as birds.
No one slips them
over children's feet.

(Michael Hamburger)

MEETING

For Michael Hamburger

Barn owl
daughter of snow,
subject to the night wind,

yet taking root
with her talons
in the rotten scab of walls,

beak face
with round eyes,
heart-rigid mask
of feathers a white fire
that touches neither time nor space.

Coldly the wind blows
against the old homestead,
in the yard pale folk,
sledges, baggage, lamps covered with snow,

in the pots death,
in the pitchers poison,
the last will nailed to a post.

The hidden thing
under the rocks' claws,
the opening into night,
the terror of death
thrust into flesh like stinging salt.

Let us go down
in the language of angels
to the broken bricks of Babel.

(Michael Hamburger)

MELPOMENE

The forest bitter, spiky,
no shore breeze, no foothills,
the grass grows matted, death will come
with horses' hooves, endlessly
over the steppes' mounds, we went back,
searching the sky for the fort
that could not be razed.

The villages hostile,
the cottages cleared out in haste,
smoked skin on the attic beams,
snare netting, bone amulets.
All over the country an evil reverence,
animals' heads in the mist, divination
by willow wands.

Later, up in the North,
stag-eyed men
rushed by on horseback.
We buried the dead.
It was hard
to break the soil with our axes,
fire had to thaw it out.

The blood of sacrificed cockerels
was not accepted.

(Michael Hamburger)

JOHANNES BOBROWSKI (1917-1965)

Born in Tilsit, East Prussia, near the Lithuanian border, Bobrowski moved
to Königsberg with his family at the age of eleven. There his studies in the
history of art were interrupted by World War II, during which he was held
as a prisoner of war in Russia. After the war he settled in East Berlin. His
small oeuvre began to bring him recognition only during the last few years
of his life; besides two novels and three collections of short fiction, his
collections of poetry are: *Sarmatische Zeit* (Sarmatian Time, 1961),
Schattenland Ströme (1962; Eng. tr. Shadow Land, 1966), *Wetterzeichen*
(Sign of Approaching Storm, 1966), and *Im Windgesträuch* (In Wind's

Underbrush, 1970). One of German literature's most significant links with
the Baltic East in our time, Bobrowski, along with Huchel and Kunert, has
also been one of the most influential East German poets in the West.

THE LATVIAN AUTUMN

The thicket of deadly nightshade
is open, he steps
into the clearing, the dance
of the hens around the birch stumps is forgotten, he walks
past the tree round which the herons flew, he has sung
in the meadows.

Oh that the swath of hay,
where he lay in the bright night,
might fly scattered by winds
on the banks—

when the river is no longer awake,
the clouds above it, voices
of birds, calls:
We shall come no more.

Then I light your light,
which I cannot see, I placed
my hands above it, close
round the flame, it stood still,
reddish in nothing but night
(like the castle which fell
in ruins over the slope,
like the little winged snake
of light through the river, like the hair
of the Jewish child)
and did not burn me.

(Ruth and Matthew Mead)

SHADOWLAND

The rustling voices,
leaves, birds, I came
three ways
before a great snow.
On the bank, burrs and awns
in her ringlets, Ragana with her hounds
shouted for the ferryman, he stood
in the water, midstream.

Once,
following the mists
across the dell with golden wings,
the bustard flew, they set
horny feet on the grass,
light, the day, flew after them.

Cold. On the tip of a grass-blade
the emptiness, white,
reaching to the sky. But the tree
old, there is

a shore, mists with thin
bones move on the river.

Darkness, whoever lives here
speaks with the bird's voice.
Lanterns have glided
above the forests.
No breath has moved them.

(Ruth and Matthew Mead)

SANCTUARY

Light, falling
with a curve
of the burdock leaf, the line of light—
wind, the glassy wing
stirs on the bank.

Come and go and come again,
come and stay, a house,
a house of mist, stands before the forest,
roofs of smoke,
towers of birdcalls,
birch branches secure the door at evening.

Restless we lie there,
a shawl of shadows on our shoulders,
the breezes move round
the fishermen's fires
with reddish fins,
you speak, alien voice,
I hear you with alien ear.

(Ruth and Matthew Mead)

PLACE OF FIRE

We saw that sky. Blackness
moved on the water, the fires
beat, darkness with trembling
lights stepped forward in front
of the wood of the bank, in animal hide.
We heard
the mouths in the foliage.

That sky stood
unmoved. And was made
of storms and tore us forward,
screaming we saw the earth
ascending with fields and rivers,
forest, the flying fires
benumbed.

The river remained deep. The pungence
of damp grass
rose. The voice of the cricket
lifted behind us, there was
a tree behind us,
the black alder.

We saw the sky that
perished in the darkness, sky
of fields and flying ancient
groves. Steps came
across the marsh, they
stamped out the fires.

(Ruth and Matthew Mead)

THE DESERTED HOUSE

The avenue
defined
by the footsteps of the dead. How the echo
descended over the sea
of air, beneath the trees
ivy creeps, the roots
show, the silence
approaches with birds, white voices.
In the house
walked shadows, a strange conversation
beneath the window. The mice
scurry

through the broken spinet.
I saw an old woman
at the end of the road
in a black shawl
on the stone
she looked southward.
Above the sand
with hard split leaves
the thistle bloomed.
There the sky was
open, in the color of a child's hair.
Beautiful earth fatherland.

(Ruth and Matthew Mead)

THE WILD SWAN

Songs of the shepherds
stretched out
on the billow of blood,
he can see you:
he who rises, the wild swan,
ocher-beaked
follows the cracking
ice. We hear him darkly,
we met the shepherds,
over the moors the birch winds
were sleeping. And quiet
was our discourse,
a fire,
fallen at night.

Rivers, we shall become of
like mind with you in the wafting
year, on the wellspring of summer,
listening: the wild swan flies off,
and about the campaign of
Igor the sorrowful lay rings,
is sung up on towers
in white,
whispering breeze.

(Juliette Victor-Rood)

THE LITHUANIAN WELL

My paths out of sand, the heavens
over the willow thicket.
Wooden bucket, travel up.
Drench me with earth.

Hours away, lark, your song,
by the head of the hawk.
If the sower hears you,
the reaper has forgotten you.

172

Glance to the plowed-up field,
the carts are coming, the wind's howl.
Ladler-maid, lean into the light.
Sing your mouth pale.

(Juliette Victor-Rood)

THE WANDERER

Evenings,
the stream now sounds,
the labored breath of forests,
heavens, flown through
by screeching birds, coasts
of darkness, old,
above them fires of the stars.

I've lived as men are wont,
forgotten to count the gates,
the open ones. On the closed ones
I have knocked.

Every gate is open.
The hailer stands with extended
arms. So step to the table.
Speak: the forests resound,
through the now breathing stream
the fish fly, the heavens
tremble with fires.

(Juliette Victor-Rood)

HANNS CIBULKA (b. 1920)

Born in the German-speaking area between Germany and Czechoslovakia, Cibulka studied library science at Berlin, and is now a librarian at Gotha. Besides two volumes of prose he has published several poetry collections, including *Zwei Silben* (Two Syllables, 1959), *Sizilianisches Tagebuch* (Sicilian Journal, 1960), *Arioso* (1967), and *Windrose* (Compass Card, 1968).

KREMENEZ

Land,
I have half-forgotten
your language,
my grammar
has gone barbaric,
but year after year,
when the December wind comes,
I return to Kremenez,
to the village
with the different sky.

There my thoughts do not
tap at the window
like a homeless bird,
there you spread for me
a table with your words,
there the grain,
even in ringing frost,
still silently grows
under the blanket
of sleep.

(Ewald Osers)

KUKS

Your glories,
Matthias Bernard Braun,
have stepped down
from their plinth,
no one knows
where
they have gone.

The angel
of annunciation
has departed,
he too
failed to leave
an address.

The days
peel away,
lime
on the tree trunks,
the terrace
before the church
is deserted.

Left behind
are a few inscriptions
chiselled
into the sandstone
plinth.

(Ewald Osers)

MATHEMATICS

"And the angel of numbers
is flying
from 1 to 2...."
Rafael Alberti

The numbers
on our childhood slates
form ranks
when the angel with the integral
enters the room

He comes with e^{-x}

with $\pi \sqrt{n}$

his tunic
is covered
with the snow of infinity

Angel of intuition
with your letters
we illumine
n-dimensional spaces
in the universe

Gone
the Euclidean tranquility
of the world

(Ewald Osers)

SKY FULL OF SWALLOWS

It's easy
to live on
the sun's shores,
my delicate
wind-swift bird.

The bright
open landscape
with the olive leaf,
that is your home,
high above the blue
bed of rivers.

What are they to you,
the proud flight
of the eagle,
the lightning
in the eyes
of the owl?

In the elm's light
of night
you return to the
village,
domestic
is what you are
called.

My smoke-blue
queen,
with you would I live
by a river
which never
darkens.

(Ewald Osers)

GÜNTER KUNERT (b. 1929)

A native of East Berlin, Günter Kunert lives at present on a three-year
passport (issued in 1979) at Itzehoe, West Germany. He is the author of
numerous books of poetry, of criticism, of radio plays, and of prose fiction.
His reputation as a poet is respected in both East and West; some of the
recent volumes on which it is based are: *Unterwegs nach Utopia* (On the
Way to Utopia, 1977), *Camera obscura* (1978), and *Abtötungsverfahren*
(Mortification Procedure, 1980). A multigenre volume, containing stories,
poems, and essays, entitled *Die Schreie der Fledermäuse* (The Cries of
Bats), appeared with the Hanser Verlag in 1979. Kunert's poetry has been

read on the BBC; he has lectured in England at the University of Warwick, and has been a visiting lecturer at the University of Texas at Austin.

EVERY DAY

A great grey swollen body
that digests all things transmutes all
Its piteous convulsions even
demanding applause

A great grey swollen body
carries out the law of transmutation
taking no notice of us

Every word turns into air
as it is spoken
Every creature into a copy of itself
saying goodbye observed
by itself
fascinated and defenseless and
every day
invariably imperceptibly
left behind

(Michael Hamburger)

VAIN ATTEMPT

On top of all earthly torments
this one: not to hit on the word
Your brain choked with dumbness
Your eyes averted and staring out:
Rows of rectangular windows
Like insoluble crossword puzzles

Involuntarily your fingers feel
the material
of which this world is made
cheap stuff

nothing that's made to last:
neither flesh nor blood
neither ankles nor arteries
not even the glaring paper
full of expectancy full of threats
to forget you
if you don't give it the word
that you lack

(Michael Hamburger)

NEWS FROM BEHIND THE SEVEN HILLS

Still letters reach us
even if we no longer know
what's written in them
Complaints about the weather
sound ominous
Every sincere greeting
a scream
Every postscript
a final farewell

How reply with what words
when all of them
only signify what they mean
and where
they're decoded
sound like a secret
turned inside out

simply incomprehensibly
simple

(Michael Hamburger)

REALIZATION

Did you know
who you are before you
stood in front of the cast-iron machine

Under the bronze worn away
greasy blackness
        bumpy curlicues of metal
here and there remains of varnish
the cut pane gently veiled
with dirt
shiny levers buttons slots
geometric lines
letters figures signs:
when the flap jumped open
renewed after long decades
when the shelf lay exposed
final and indifferent
you knew who you are:
When you had removed the lightning
you turned away
for good really
and from now on for ever

(Michael Hamburger)

IMPLANT

You only have to be born
and soon you receive a visit:
the bogeyman
gets inside you
right up to the saliva mark
right up to your teeth
He makes your fingers dance
He bleaches your pelt
He extinguishes your thoughts
He throttles your voice:
Speak as though begging for mercy:
In any case every word is a lie
It betrays you

What are you saying what have you said
that you ought not to have said?
Only those graced with dumbness
remain strangers to themselves

Your doppelgänger though
this spongy figure inside your skin
penetrates your pores
as visible evidence
of his presence

(Michael Hamburger)

REINER KUNZE (b. 1933)

Of humble origins—the son of a miner—Kunze has both studied and
taught at the Karl Marx University in Leipzig. In 1961 and 1962 he lived in
Czechoslovakia, has translated poetry and prose from the Czech, and
remains in close touch with Czech writers. His minimalist poetry has been
of seminal influence in the West. Since 1955 his poetry collections have
been *Die Zukunft sitzt am Tische* (The Future Sits at the Table; with Egon
Günther, 1955), *Vögel über dem Tau* (Birds over the Dew, 1959),
Widmungen (Dedications, 1963), *Sensible Wege* (Sensible Ways, 1969)
and *Zimmerlautstärke* (With the Volume Turned Down, 1972), the last-
named published in West Germany. In 1973 an East German selection of
his verse appeared under the title *Brief mit blauem Siegel* (Letter with a
Blue Seal). His latest book, *Die wunderbaren Jahre* (The Wonderful
Years, 1976), became a best seller in West Germany. Kunze is also the
recipient of numerous prizes, including the Literature Award of the
Bavarian Academy of Arts (1973) and the Georg Büchner Prize (1977).

REFUGE EVEN BEYOND REFUGE

(for Peter Huchel)

Here only the wind enters unbidden

Here
only God calls

Countless circuits are laid at his behest
from heaven to earth

From the roof of the empty cowshed
to the roof the empty sheep pen
shrilling from wooden gutter
the rain jet

What are you doing, asks God

Lord, I say, it is
raining, what is there
to do

And his answer sprouts
green through all windows

(Ewald Osers)

REPLY

My father, you say,
my father down the pit
has gashes in his back,
scars,
        traces of fallen rock,
while I, yes I,
sing of love.

I say:
yes, that's why.

(Ewald Osers)

HYMN TO A WOMAN UNDER INTERROGATION

Bad (she said) was
the moment of
undressing

Then
exposed to their gaze she
discovered everything

about them

(Ewald Osers)

ALMOST A SPRING POEM

Birds, postilions, when
you start singing the letter
with the blue seal will arrive, with stamps
bursting into flower with words
reading:

Nothing
lasts
forever

(Ewald Osers)

A KIND OF HOPE

A grave in the earth

Hope of resurrection
in a blade of grass

(No grave slab:

An end at last to
being thwarted by stone)

(Ewald Osers)

MEDITATION

You wonder what that is, daughter.

Still sitting at the desk toward morning,
a moth
sleeping
on the pantleg

and neither knows of the other

(Lori M. Fisher)

RETURN FROM THE CONVENTION

Close the forest behind you, the door
filled with song

The wild game
break out of their blackness at night

The rustling of the spruces in the ear: the tape that
shrills in the head is

erased

(Lori M. Fisher)

DIARY PAGE 69

Kottenheide where you
dream where thoughts
are fertilized each morning

Where you let them ripen

If the rain
doesn't hang suspended in the forest; if it
doesn't hail pigeon eggs
in June

If the axe
doesn't strike before the tree
spreads its seeds

(Lori M. Fisher)

HEINZ CZECHOWSKI (b. 1935)

Born in Dresden, Czechowski worked as an industrial draftsman and a commercial artist before turning to a career in writing. Between 1958 and 1961 he studied literature at Leipzig, then worked in publishing at Halle, where he now resides. His three collections of verse, *Nachmittag eines Liebespaares* (Afternoon of a Loving Pair, 1962), *Wasserfahrt* (Journey by Water, 1967), and *Schafe und Sterne* (Sheep and Stars, 1974), were followed in 1972 by a collection of essays, *Spruch und Widerspruch* (Dictum and Contradiction).

WALK THROUGH THE CITY

1

The river drowns itself
in the surfeit of its water.
Above me shines
the orb of the great Inventor:
Let him buy the whole
city from us, what
do I care?

Poets inhabit it,
actors, successful
engineers, one
who once was my friend
now talks to himself:
no one
listens to him.

2

To write poems is difficult.
What should I write poems about? The trees
that stand on the hillsides,
the smiling landscape,
the gentle slopes? Here
I should like to stay.

3

The streets,
desecrated with corpses,
are slowly reviving:
a cross for the dead, up there
behind the forest looking down on the city,
for the nameless.

4

I too am nameless
but alive.
Always the dead
hold me here,
no neon lights, no
neo-baroque.

Ah, fellow-citizens,
to be honest and
free from pretense, thus
between life and death
live one's life, not
for the all-powerful conveyor belt.

5

Still
the Kings are dancing
behind the trimmed hedges: skeletons
stripped of their flesh by
the X-ray of time. Gone
the amorous chase under plane-trees,
the pouncing of
flesh upon flesh.

City, island of citizens,
not for history
do we live here. You
sent me away, I
come back.
Unnamed
I do not tire
of giving you names: your contradiction
which lurks in the stones
that have left me.

(Ewald Osers)

SARAH KIRSCH (b. 1935)

Sarah Kirsch, one of the most popular East German poets writing now, was born in Limlingerode, in the Harz region. She studied biology at Halle, and from 1963 until 1965 was enrolled at the Johannes R. Becher Literary Institute in Leipzig. Between 1965 and 1977 she lived as a freelance writer in Halle and East Berlin, and also traveled widely in the Soviet Union and in East bloc countries. Since 1977 she has lived in West Berlin. Besides voluminous translation activity (Tsvetaeva and Akhmatova), short fiction, and prose portraits of women writers, Kirsch has to date published ten books of poetry, among them *Zaubersprüche* (Magical Incantations, 1973) and *Rückenwind* (Back Wind, 1976), the sources of some of the translations offered here.

DON JUAN IS COMING IN THE MORNING

Don Juan is coming in the morning
so he wrote in his telegram which
made me consider I was counting on the moon
and fountains there wasn't much time left
to paint my eyes
bigger to wash my feet
I stood where the city begins saw him
approaching on a racing bike
his coat flapping the white scarf

fluttering over his shoulder
his lips cracked his eyes set deep
I asked him why he was so early
probably has a rendezvous with a beauty later
don't be silly! he stood his bike up
at an angle he took his hat off
laid us both in the grass
beginning to grow lush all around us
wound up metal birds they began singing
Variations on a Theme by Mozart
resounded I know that piece
he said and all about stereo components
Schönberg and
I'll do you'll see now it'll be good

(Tanaquil Taubes)

SEVEN SKINS

The onion lies peeled and white on the cold stove.
It glows from its innermost skin alongside the knife.
The onion alone. The knife alone. The housewife
ran down the stairs crying—she was so affected
by the onion or by the position of the sun over the
adjacent house.
Unless she comes back, unless she comes back soon,
the husband will find the onion mild and
      the knife misted over.

(Tanaquil Taubes)

DRINKING VODKA

Brown birds brown leaves
soft resin globe on the bark
of singly large trees
whose tops stars ornament

Before the castle old marble-men
stretch their beautiful limbs

blackbirds nested in the lap
of a white fountain-lady

Inside in the huge chairs
lions' heads on their arms
we sit and drink vodka
making of our throats a wilderness:

Red flower-clusters tumble
on the silver-white river
and by moonlight the curving crabs
dig themselves into the ground

until our gullets harden
until the great liquid streams
roll softly down inside us
and we grow increasingly solemn

while the others all shout
and link their arms
exchange the toast of friendship
which tomorrow they won't remember

Come let's out into the cold.
In the darkness of the moon
we'll walk on polished roads
that bend around the mountains.

(Wayne Kvam)

STOPOVER

When it comes to Christmas time
the poets return
to their diligent wives
Oh where they haven't been
for the entire year
what they haven't heard haven't
considered, written in their newspaper
clambering through factories, they
imparted social graces to the potatoes, inspected
the smoke that creeps and climbs

they've swallowed everything sometimes Manhattan-
cocktails only for the name, they intensified
the class struggle meditated
on the abstract while fishing, until one day
through their thin jackets the cold comes
Longing
for a real fish in the pan
suddenly overwhelms them and remembrance
of the woman who's warmed herself by the fire
then anger
is left behind in the large cities, they come
with strange hats for their children
even do the laundry play the piano, until
after New Year's they've had enough, then
pick a fight, depart relieved
in gloves from under the Christmas tree.

(Wayne Kvam)

THE AIR ALREADY SMELLS LIKE SNOW

The air already smells like snow, my lover
Wears long hair, ah the winter, the winter that
Throws us closely together approaches, comes on
With the greyhounds' speed. Ice-flowers
He scatters on the window, the coals glow in the stove, and
You beautiful snow-white one lay your head in my
    Lap
I say that is
The sleigh that no longer stops, snow falls
In our hearts' middle, it glows
On the ash-bins in the yard Darling whispers the blackbird.

(Wayne Kvam)

DATE

He came on the 28th of February, stood
Before my window in a bearskin coat said
Oh how dizzy I am. To this height

I could accustom you, beautiful one
Learn to carry me and I'll
Make myself light. And for you
Many a miracle shall transpire: my hair
Will grow through your fingers your mouth
Become the imprint of mine you'll hear me
When I'm not there. Speak my name
Out into the winds: all will happen.

Lovely one shall we be Juliet and Romeo?
The circumstances
Are favorable, we live indeed
In the same city, but the countries
Our registered countries act up, mine
Holds me and holds me it hangs on me so we
Could be very unhappy ah you were speaking
To me just now.

(Wayne Kvam)

KURT BARTSCH (b. 1937)

A Berliner, Bartsch lives in East Berlin. He has worked at odd jobs, including those of clerk, telephone operator, and truck driver's mate, preceding his studies in literature at Leipzig. He is the author of three books of verse: *Poesiealbum* (Poetry Album), *Zugluft* (Draft of Air) (both 1968), and *Lachmaschine* (Laugh Machine, 1971), the last-named published in West Berlin.

RITA

Rita's husband gave Rita her walking papers
Now Rita sits there and bends an elbow
In the bar Old Schönhauser Street
The building is condemned, but the
Business still flourishes. Waiter, two brandies! This is
My treat, says Rita, because Rita's husband
Gave Rita her walking papers, Rita drinks. What should I do
I'm over forty, divorced, three kids on my back
The youngest is thirteen and goes with men already

Practices early, when *I* was thirteen
Everything was different, says Rita. First war, then the hunger.
At sixteen, she says, I pounded stones
We dug the city out with bare hands
Now the city stands. I'm a ruin.

(Wayne Kvam)

STILL LIFE WITH CLEANING LADY

Grass grows over the dead rails
Here the streetcar once ran
From West to East East to West
CLOSE THE DOORS STAND BACK
The tracks rust, the bridges
Cross over dead streets, the streets
Obstructed by walls, concrete, so that
I don't go running off, says Trude. Ah
So much expense for one who can
Do nothing but sweep stairways, scrub floors
She smiles, then one has
To stay here, don't you think

(Wayne Kvam)

DEPOSITION

You can do what you want
I can't remember anything more.
My head is a sieve. I know only
That I know nothing that you don't know.
After all, what do you have your people for.
That would've really been funny if you hadn't known
Why it was funny in the first place.
A joke. I often tell jokes
Among a few friends, that doesn't mean anything.
Jokes are an outlet, one has
To let off steam now and then.
Have you heard this one? Stalin enters into heaven.
I knew that you'd heard it. Well

192

One can't help that. In case you
Hand over my things now
Please don't forget the rope. I am
As you know, a rope-dancer by trade.
Since when? Since five-hundred years ago.
My name is Eulenspiegel. No,
The way it sounds. Eulenspiegel. I'll
spell it.

(Wayne Kvam)

BERND JENTZSCH (b. 1940)

Born in Plauen, Jentzsch pursued studies in German literature and in the
history of art at Jena and at Leipzig. Since 1965 he has worked in
publishing in East Berlin. His early volume of verse, *Alphabet des Morgens*
(Alphabet of Morning), dates from 1961; he is also the author of a volume
of prose, *Jungfer im Grünen* (Young Woman in Green), published in West
Germany in 1973. The bulk of his poetry has appeared in journals and in
anthologies. Jentzsch has also been active as an editor of anthologies and as
a translator from a number of European languages.

TEREZIN, THE GRAVEYARD

For Robert Desnos

Provided
that the occupants of the cemetery keep calm
and do not burst open those thirty inches
of alien soil, but
sleep on and continue to view the luxuriant
ivy from below. Provided
that the victims, recklessly declared immortal,
will not take us, take you, take me
at our word, and one fine judgment day appear
at the door, casually,
clearing their throats. Assuming
that the crumbling coffins do not rise up
with what provides street names, calls for costly
wreaths and, so it is claimed, is gradually

losing emotive appeal. Provided therefore
that the eulogized occupants of the cemetery,
faced with the spotless sightseeing buses,
keep perfectly calm,
I scream:
That man,
seize him by
hands and feet. One man,
and that's one too many, one man
is standing here and is not
moved, he has, generally speaking,
positively forgotten
everything.

(Ewald Osers)

THE DEAD FATHER

In the end he was light enough.
He carried nothing upon him.
No place was assigned to him.
No objective crowned his advancement.
Out of all-sealing time
he floated away,
girdled with memories.

(Ewald Osers)

LUTZ RATHENOW (b. 1952)

Lutz Rathenow studied literature and history at the University of Jena; he moved to East Berlin in 1977 and there gradually established himself as a freelance writer. He was among the founders of a literary group that sponsored readings of many leading East German writers in private homes. In 1980 he published a book of short stories with the Ullstein Verlag, entitled *Mit dem Schlimmsten wurde schon gerechnet* (We Are Prepared for the Worst). His prose is often reminiscent of Kafka, confronting us with people possessed of distorted senses of reality. His poetry often celebrates brief moments of peace in a generally threatening world.

PRAGUE

Why just now a visit
in the middle of Winter
to this city
that I hardly know
That has promised itself
to a frost which seems never to tire:
each mouth hastily empties
mutilated sentences and runs
down the street with its owner
to a warmer place
But already this pleasant feeling
in a moving train
Thoughts are in motion
and spring from the rails
A border is crossed
without effort
Give yourself to a dream
and still move on—
only this certainty
that comforts
could be a reason
And there is something in this city
whose touch I fear
and seek: the train station the collection
of streets and towers
something that makes me wish
to come again
A cup of coffee in the cafe
The waiting for a friend
who doesn't come
The conversation of two strangers
Reading Jiří Wolker
Tormented by pain I break
your pair of eyes in pieces
Someone rubs on the frozen window
And outside the sighs from steps
in snow revive
One pulls the scarf tightly
about the face and hurries
to cross the tracks
The sound of a record
that never seems to end

Why just now in the middle of Winter
a poem on this city
that I have not visited
for years

(Boria Sax)

NO PICTURE

A knife
That has learned its trade again
Blood—say a pitcher full—
Chased by its loneliness
An anger
That was quickly past
A piece of field a piece of man
Reluctantly left behind

(Boria Sax)

AFTER DAVID SAMUILOW

The poem: modest
as though it were a knotty branch.
On the path where I walk
it serves me as a support.
And, to scare away the dogs,
to keep time as I stroll.
It's bare, without decoration—
without ornament (big deal).
It serves me as a support
and it's also good for blows.

(Boria Sax)

CZECHOSLOVAK

INTRODUCTION TO CZECHOSLOVAK POETRY

The death of Vitězslav Nezval in April of 1958 left modern Czech poetry bereft of one of its founders and most essential early voices. Soon after World War I Nezval and Karel Teige together founded poetism, Czech poetry's answer to surrealism, and their co-authored theoretical document appeared in the spring of 1924, six months before the publication of André Breton's first surrealist manifesto. Nezval's verse was itself the best embodiment of this poetic of playful and free association, and of childlike wonder at the beauty of the everyday. In spirit (if not in minutiae of approach) and above all in the choice of a poetic locus Nezval was joined by Jaroslav Seifert and František Hrubín, and these three avatars of modernism bestowed upon European letters a body of urban poetry—a celebration of the city of Prague—paralleled at least in this anthology. Even more than Nezval, whose *Prague with Fingers of Rain* (1936) now seems somewhat dated, Seifert and Hrubín suggest in their own Prague poetry Mandelstam's doctrine that poetry is architecture and that the poetic word is its building block.

How varied Czech poetry was between the two world wars, and how the wealth versus poverty of the city affected poets' visions, we are not able to show here. An entire now submerged generation counts among its members the proletarian poets S.K. Neumann (1875-1947) and Jiří Wolker (1900-1924), the vitalist Josef Hora (1891-1945) and, most significantly, the complex, tragically attuned František Halas (1901-1949). Even among the living poets we could not include sincere and tenacious workers in poetry such as Miroslav Florian (b. 1931), Jiří Kolář (b. 1914), Jan Skácel (b. 1922), or Ivan Wernisch (b. 1942). The gifts that we have received, however, do enable us to suggest an echo of the almost legendary complexity and variety, in concerns and voices, in Czech poetry since the prewar years. In fact we must go back as far as Valdimír Holan in order to gain a sense of how true it is that not all of Prague's poets were preoccupied with their city alone. Holan educates his poetic being in a school of intimate feelings, and his persona travels to realms that lie beyond physical topography. "Cave of words! Only a real poet returns from its silence, / so that, already old, he may find a crying child, / abandoned by the world at its threshold." If there is return, albeit tenuous, to a sense of metropolitan presence along with one of emotional privacy, it is in the work of one of two poets born during the twenties, Antonín Bartušek. In his title poem "Royal Progress" there is a texture of autumn, of yellowing foliage and dampness, of bridge and castle, shading over into a surface of self: "At dawn we trip over our own shadow."

How does a poet dig himself out of what Bartušek calls "the avalanche of home"? One way is by being a scientist. The analytical mind of Miroslav Holub, a research biologist, is on duty performing its precision work in the laboratory of poetry. There are prose poems, recording perceptions either in the real laboratory or on journeys, to Philadelphia or to New York. By mild contrast, many of the poems in *Sagittal Section* are meditative and dissective at once. Cosmic "Crocheting" does not lie far from the mythological intimations of "The Garden of Old People." But the real surprises that the synthesis of poetry and science holds for a poet of Holub's fine dryness are those of language, the scientific tool. If this again disquiets, if it somehow once again seems to gainsay our humanity, we need but ask what a poet makes of his linguistic opportunities. Holub's language, especially in the cycle of "Brief Reflections," the source of the Charlemagne poem, is fresh and precise. Josef Hanzlík, fifteen years Holub's junior, is preoccupied by thoughts of death and of escape from this "vale of tears." In "Three Cheers for Herod" (printed entire in the Penguin *Three Czech Poets,* pp. 129-32) the heaven-dwelling children slaughtered by the king gather to pay him tribute for his having saved them from the sorrows of this life. Affirmation of sorts follows later: "Who's that driving an empty cart / who but the gravedigger who like the captain / dies last." Like Holub, Hanzlík is open to the tonalities of irony although, paradoxically, he seems to question the possibility of distance from the envisioned subject.

The youngest poet in our group of Czechs, Antonín Brousek, tosses the coin; down it comes on the side of his contemporary. Sorrow and parting are valid diapasons in the ever-moving music of contemporary lyricism. *Pillars of Salt* is a poetry of parting, from the city from where friends and associates have parted before the poet: "I would have left long ago / had I not been detained by words / which it hurts me to give up, / cumbersome luggage / above the permitted weight." Motionless and moving, the pillars throughout this long poem remain; the salt of speech goes on stinging.

Of the two Slovak writers whose work we were able to include the older, Miroslav Válek, is represented by three very different poems. "Steel-Rod Benders" seems central in the vision of a craft-conscious poet: "bending steel rods. / It is, in a way, poetic work. / The iron, like a verse, resists the human will." Is there something universal about poetic closures and their limitations? "If I'd had bigger muscles I'd have become a steel-rod bender." "One could do worse than be a swinger of birches." But it does not take uncovering a perhaps unwitting allusion to discover affinities among neighbors. Brousek and Válek are both fascinated with apples, with the iron content of the fruit, with iron and steel. Certainly they both portray possibilities of parting. For a sharp poet like Válek what better sense of return following departure than the poetic report on the "abnormal sense of hearing" of a man "deaf as a post"? In ages "modern" and "postmodern" the symbols of our frightening being surround us. Whether here or there. Take the warm, personal lyrics of Daniel Simko, a poet for whom departure is accomplished fact. Somewhat on the model of Brodsky's "Elegy: for Robert Lowell"—or rather, on that precedent—Simko appears here with poems he wrote in English. The effect of suggestion is strong. In some poems ("Thinking of My Father on a Bus to Baltimore") his voice is akin to that in the America poems of Holub (which he has translated); as we progress through this all too modest group we overhear a European coming back—to himself ("Bells. You are alone."). And then we come to the concluding piece, "The Jewish Cemetery in Prague." We have come full circle. We can leave again. The town "has stepped out of your bodies."

JAROSLAV SEIFERT (1901–1986)

Jaroslav Seifert is the dean of living Czech poets. He was born in Prague, worked for a time as a journalist, and has for many years made his living as a freelance writer. One of the most prolific Czech writers of this century, Seifert is the author of over thirty volumes of poetry between 1921 and 1970, along with a number of published selections of his verse. The poems translated here are from Seifert's most recent (and perhaps last) poetry collection entitled *Deštnik z Piccadilly* (Umbrella from Piccadilly, 1979).

A GARLAND ON THE WRIST

I too, on Corpus Christi Day,
used to inhale the fragrance of incense
and thread a garland on my wrist
of fresh spring flowers.
I too used to gaze up towards the sky
devoutly, listening to the bells.
I thought that was enough
but it wasn't.

How often has a fugitive spring
with its heel stirred up a flurry of blossom
under my window,
and long ago I realized
that a fragrant bloom

and a woman's body radiating nakedness
are two things
more beautiful than anything else
on this miserable earth.

Bloom and bloom,
two blooms so close to each other.

But life escaped me hurriedly
like water through my fingers,
even before
I'd managed to assuage my thirst...

Where are those garlands of spring blossom!
Today, as I hear the creak
of the death chamber's door,
when I've nothing left to believe in
except something too much akin to nothing,
when the blood is pounding in my veins
like the condemned man's drum,
when all that remains is the stereotype
and all hope is as worthless
as the old collar
of a dead and mangy dog,
I sleep badly at night.

And that is how I heard
someone tap softly
on my half-closed window.

It was just a branch of the tree
flowering in spring,
and my two sticks
with which I drag myself about
day after day
did not for once have to transform themselves
into a pair of wings.

(Ewald Osers)

FINGERPRINTS

Even by force I make the night surrender
pleasant dreams.
Alas, mostly in vain.
But life, at least, allows us
to return against the current of time
not without vertigo, but with some slight regret
and a tear of sadness,
all the way
to where our memory reaches.

Remembrances, however, have a woman's skin.
When you taste them with the tip of your tongue
they taste sweet
and have an exciting fragrance.
So what!

The statue of the Vltava river by Václav Pachner
in the facade of the Clam-Gallas palace
pours from its jug
a stream of water
intertwined with stars.
She has long bewitched my eyes
with her shapely nudity.

Confused they strayed a long time
over her body,
not knowing where to settle first.
On her delightful face
or on the virgin charms
of her lily-of-the-valley breasts
which so often are the crown
of all the beauty of the female form
in all parts of the world.

I must have been fourteen,
or maybe a year older,
as I stood there bewildered
as if waiting
for her to raise her eyes to me
and smile.

One moment when I thought
no one was watching
I managed, from the basin's edge,
to clasp her leg in my palm.
Higher I didn't get.
It was rough, it was made of sandstone,
and it was cold.
It was still snowing lightly.
But a hot wave of desire
like an electric shock
surged through my blood.

But if love is something more
than a mere touch,
and it is that,
a drop of dew can sometimes be enough,
suddenly trickling down onto your hand
from a flower petal.
Your head suddenly spins
as if your thirsty lips had gulped
some heavy wine.

In the doorway of the Clementinum nearby
stood a policeman.
And while the wind was ruffling
the cock's tail feather on his hat
he looked around.

He could have easily run me in.
No doubt I'd left my fingerprints
on the girl's calf.
Perhaps I had committed an offense
against public morality,
I don't know!
I know nothing about the law.
Yet I was sentenced after all
to lifelong punishment.

If love is a labyrinth
full of glittering mirrors,
and it is that,
I'd crossed its threshold
and entered.

And from the bewitching glitter of mirrors
I haven't found the way out
to this day.

(Ewald Osers)

VLADIMÍR HOLAN (1905-1980)

Vladimír Holan was born in Prague, and spent most of his life there. From 1933 until 1938 he was editor of the journal Život (Life), but since 1940 has devoted his time entirely to his poetry. From introspection and philosophical meditation in the thirties, exemplified in such volumes as *Triumf smrti* (The Triumph of Death, 1930) and *Oblouk* (The Arch, 1934), Holan went on to developing an art characterized by strong intellectual and emotional tension, and by expressionistic form. During the sixties, when Holan's influence among young poets was strongest, his poetry also reached its high point and he published his finest collections, including *Bez názvu* (Untitled, 1963), *Na postupu* (On the Advance, 1964), the meditative dialogue *Noc s Hamletem* (A Night with Hamlet, published the same year) and, most important, his "poetic diary" from the years 1949-1955, *Bolest* (Pain, 1965). A definitive edition of *Bolest* appeared the year following. The poems included here are from this volume, as well as from *Bez názvu*.

AT MOTHER'S AFTER MANY YEARS

This is the time when the fire on the hearth
needs to be banked with ashes...
Your old mother's hands will do it,
hands which tremble, but hands
whose trembling is still the measure
of reassurance... Having been lulled by them you fall asleep
and it feels so good... Custom, warmth, delight and calm,
the breath's intimacy with something animal-heavenly,
to be both giver and receiver
when you lose yourself:
all this denies that you could be over forty.

205

And indeed, if you should sob a little towards morning,
then it is only because
a child never laughs in his sleep,
but only cries... A child!

(Bronislava Volek and Andrew Durkin)

THE CAVE OF WORDS

Not with impunity the youth enters, light in hand,
the cave of words... Bold, he hardly suspects
where it is he finds himself... Young, though suffering,
he does not know what pain is... A master before his time,
he will escape without having entered,
and he will blame a century not yet of age...

The cave of words!...
Only a real poet and at his own risk
will ruin his wings in madness there and lose the knowledge
of how to return them to the earth's gravity
yet not to hurt her, who attracts the earth...

Cave of words! Only a real poet returns from its silence,
so that, already old, he may find a crying child,
abandoned by the world at its threshold...

(Bronislava Volek and Andrew Durkin)

MEMORY II
For František Tichý

We looked for hours everywhere
in vain for pimpernel. We came out of the groves
and just at midday we found ourselves standing on a heath.
The air had dried into sheet metal. We watched
the opposite hillside, thickly overgrown
with varied trees and bushes. They were stiff as we were.
I was just about to ask something
when in that immobile, fixed mass,
bewitched to the point of chill, a single tree,

standing in a single place,
suddenly began to shiver
like a hexachord, but without a sound.
You would have said it was rejoicing from the heart's lightness
and so, even adventure itself.
But then the tree began to rustle
as silver rustles because it will blacken.
But then the tree began to shiver,
as the skirt of a woman shivers when it brushes against
a man's clothes during the reading of a book in a madhouse.
But then the tree began to shake and toss,
but as if it were being shaken and tossed by someone
who looks into the black-eyed pit of love—
and I felt as if I should die right now...
"Don't be afraid," said my father, "it's an aspen!"
But even today I remember how he turned pale
when later we came there
and beheld an empty chair beneath it...

(Bronislava Volek and Andrew Durkin)

From UNTITLED

1.
And Beginning There Is Not

The charnel thickens into a shadow...The boys, meanwhile,
Knock down plums, throwing bones.
This double fall and second-primal sunstroke
For a spell blinks with the ear of dizziness
And afterwards dies out all colors
In the cemetery's negritude.

And beginning there is not. Only the first and second end together
Cough up the blood of music
On the lines of palms which feature
One year's span merely
Behind the cracks of walls.

6.
A Picture Which Is No Abyss

Under a lime tree, the saint's beard being dyed...

The unknowing air is trembling
With virginity in the graveyard's ears,
Redness becomes without itself
Of nakedness lets go the maiden's laughter—
The aroma,
Shining only when it stops to be
Denies that song
Had a man with a rock exclusively.

And still, still a picture
Which over here is no abyss,
A sign is unable to be.

10.
Over Shelley's Letters

Bliss isn't in a heart that bursts with fullness...

For, exactly at the moment when
The smoke range of pleasures sounds even in the cavernous
Reflections of sadness,
Thought wishes a sepulchre on pillars.

If water is truly flowing,
Then reconciling and free tranquility
Hardly can be in repose.

And the poet? Behold, he goes,
Goes like something that no longer is, elsewhere,
If he's to be still here...

17.
Snake

Along a rocky path there walks a man.
In his palm, several vipers...
The faintest hiss, a bit in whorls convoluting,
Shakes up the air as well as kindred
Minutes that go off, shooting
Into the glitter of cerebral afterlife,
At which the sun constricts its
Pupils of sentiment...

Not otherwise through its mourning energy
The spinal cord snake always persuades the woman,
And passion only craves an instant of extension
As if wishing hatred to remain forever, even
Posthumously—

Who penetrates, imprisons... Imprisons himself...
The boulders of barrows
To the egg are returned by gradual plasticity,
Within which waits a secret hearing only, until the world
Fails to endure through the morrow of delusions.

The sound of the man's step died... But the remaining
Unqualified menacing, gaping open
Into a more acceptable trembling,
In fact has condemned already...

Are we in apostasy or cast off to languish?

65.
Somnia et noctium fantasmata

This is the night when through waking's secret passage
Enter phantoms, all too created,
To extinction to capitulate.
Bitterly the stair step creaks and gowns are in suspense
With burnt-out concentration which will spark again
In an owl's schedule,
Somehow a feature makes the vain attempt
To be a smile and to keep the heart unaltered,

209

Some hand is there which wonders about echoes
A voice combed down over the ears of hell's anteroom,
Some steps there are, ringing only as a wavering
(For a must to them the future represents
And yet, this future does reject them)—
When, one sudden shadow bends insinuatingly
To burn away the scabs of stigmas
In the Eternal Lamp's flamelets,
Till every eye does follow him...

But you sleep on, sleep on almost willfully,
As if you were to intuit
Thanks to hairy endurance in the coffin
What is an angel's ruined crest.

(Miroslav Hanák)

FRANTIŠEK HRUBÍN (1910-1971)

The career of František Hrubín began in the "golden age" of Prague surrealism in the thirties, came to maturity in the underground volumes he published during the Nazi era, which included *Země sudička* (Land of Fate, first published in 1941), and reached its zenith in the official public-performance poems of the late forties and early fifties (among them the dramatic workers'-cabaret poems "Hiroshima" and "Nights of Job"). After this he seems to have fallen into some disfavor and thus turned to children's books, but by 1962 Hrubín had also published over a dozen volumes of poetry, as well as an autobiography. He is a master of tight verse structures, especially the sonnet, voiced with a verbal power and metaphysical impulse unique among his peers. Our selection from Hrubín's oeuvre concentrates on this phase of his work—period pieces that have a strange power to endure and seem even topical.

From THE CITY IN THE FULL MOON
for Bedřich Fučík

1.

When the great wind blew the streetlights out
pedestrians were thrust into their shadows,
cobweb faces sat within the houses
oddly filling the old familiar seats,

and only half the ancient city rested,
others groped as if they felt for bottles
in cellars, while the wind was only settled
gradually after many days and nights,

great fields of snow encircled everywhere
the city, country-places were exposed
unready, much old paper set ablaze,

and the booming wind, unleashed and sheer,
drove down the nights and mornings with its snow
in all the streets where my lost face might go.

6.

Translucent paper you, you evil woman,
unfolding quietly above the city,
you shape its life expertly, to distort
its memory, denying the strong old land—

and iron bells are to be hung again
in ancient towers long since pulled to ground,
and ashen ferry-boats, like dreams, pass down
the river to the shores of death to which you cling—

the shores of death, where last heart-beats are held
the bells still sound, the dead hands shimmer white
like trembling leaves of ice on which you light,

thus my night flies. Oh burden: to speak aloud
your other names: Moon, Death, Woman—
upon my face in place of sleep, you burn.

8.

But death has carved my brow more deeply still—
Upon his bed a sleeper turns, and hears
the quarter hours peel from all the spires,
while down upon his head dread horrors spill—

Oh mother of sleep, sleep, endless sleep,
Mother! you call the child to the childless—
great hag of rest! to sink into your softness,
and to let your arms embrace, deep, deep—

not day? And still are splendid serenades
which frost sings down upon the strings of the roads,
while the homeland voice goes searching like lost birds—

oh deep within that voice, death fades—
But death's embrace, both pain and frost now share,
sleep city, city of imperial years.

13.

Silence—how much the earth desires to tell—
translucent silence in whose starlit hour
you grow hard, you city sleeping still,
hard like boulders tempered in the flood's full power.

The smashing glasses in the wine-vaults ring
like laughing women dancing dishabille;
upon the land time casts its unleashed swell
of devastating wind, and on it brings

a storm of heavy stones until it weeps—
stones from tombs with names long since effaced—
a land whose dead oblivion erased

the vineyards' blush—Oh Lord, let us yet reap
harvests of blood, for shame has made us weak;
yet for its primal roots, the country seeks.

(Don Mager)

THE HONEYCOMB OF BEES
In Memory of Jan Novotný from Lešan
For Professor František Novotný

I pray above your grave. The two small hills
close in your peace, the field where you rest.
The clay weighs you down, my shadow falls
across you, do you sense now your own past?

And can you feel the stand of hardwood trees
(their beams enclose you), can you hear the wind
which froze your face, scraping the old larch tree
like ancient masts. The crops are gathered in.

I pray above your grave; grandson; man.
And your quicksilver touch has left my brow
as lion heads on coins wear down, and now

I barter for the sun. The smell of grain
encircles me within your fields of vast
and silent grief; and my strong grief gets lost.

To live beside you backward from the grave
I hear your footsteps stride beside my own
and when I rest in grass along the way
your body is imprinted on the lawn,

and when I brush the ripened corn, some person
seems to spray my hands with streams of grain,
and from the Commons Land a blackbird fiercely
whistles, wrapping you within my dreams.

Above the orchard, ancient bees still swarm
about their ancient queen upon her stalk
where she has joined the heights and depths of blackened

skies—my heart's disquieted alarms
are trembled by you; in my dreams you poke
your hand from churning clouds of beehive smoke.

And now you sleep in the narrow bed of death;
snowstorms of faces, memories of kin
are quick to melt your face into the earth—
you drank up each tomorrow, husbandman!

And so your heavy breath would come from distant
fields, and like a youth your flail beat
the air to fill and burst the barn's own heart,
twice each summer bringing harvest fruit.

The dormer window opened its blind eye,
a thousand splendid sparks broke from the hay,
but when the purple skies enclosed your day

you were like well-used haycarts set aside
or the salted bread stuck out of view:
The twilight passes from the fields of dew.

Through the ash trees and the hours, hums
the soundless sound of stars, while time takes measure
of our blood—the moon weighs out its beams
and presses towards the chamber's twilit door,

and sternly, sternly only stout-hearts sleep,
and sweetly, sweetly children in their gowns;
but who is free of dread to turn his step,
or break the vigil of his glass of wine?

My sleep is brief, and I commune with you;
I gulp my wine—both sleep and watch I fear—
a chicken pecks the hour's heart we share,

and I must reckon what this night can do
besides unspin the thread of life with guilt
far back into a great-grandmother's heart.

Holy John keeps watch above our house
while arsonists set blazes in the hay
and he looks down and sees our harvest stores,
so richly gathered in, now laid to waste.

All the region shuns us as they group
on Sundays at the church; they wait beneath
the solitary mid-day sun for troops
whose marching boots can quake the very earth;

behind them while they watch, the organ breaks
its thunder loose, the altar cloth is rent,
the poor folks give to crowds of beggars bent

around the door what wealthy people took;
and hearts, which shafts of whitest fury seize,
must hide in blossoms on the linden trees.

(Don Mager)

ANTONÍN BARTUŠEK (b. 1921)

Born in 1921, Bartušek studied history, history of art, and philosophy at
Charles University. Since 1950, when he took his doctorate, he has held a
variety of positions, most recently with the State Institute for the
Preservation of Historical Monuments. He has seven books of poetry to his
credit, the most recent *Antistar* (1969) and *Královská Procházka* (Royal
Progress, 1970). He is also a translator of American and European poetry
and the author of essays on art history, scenography, and literature.

HELIOTROPISM

Heliotropism
of branches, now darkened.
Ranks of shadows on the battlefield
of mist.

How my land
has changed. Golden dust
of sun in the mud, flowers under shrouds.
The white unicorn

215

all grey.
And the stars, the stars
out of sight.
Only the trees

are still fighting. Overhead
the armies of darkness
blocking the view
of the bare branches.

(Ewald Osers)

HOME

Steam
breathing from burnt saucepans
now half cooled.
In the twisting

distances the sun's blade
being dimmed, the flight
of birds taut as a bowstring,
from fields a perfume pressed

between two pages
of a closed glance. The challenge
of sunset
to the cloud

at the road's bend. How many times
have I crashed
on the stone
in the dark layers of midnight.

But you, dead
in the avalanche of home
when the bird forgot itself
with a star under last year's sky.

(Ewald Osers)

THE STRANGLED
for Ewald Osers

On a thin rope
quickly tied in the morning
round our necks
we spend the day
full of the joys of strangulation.

A chilly day,
long-redeemed flowers,
frost showing its teeth,
golden leaves
turned to mud
in the garbage pit of memories.

We are still alive,
if someone looked in
from the far side of the window
he would see a movement
as of a hanged man
slowly cooling and the branch
still gently swaying.

(Ewald Osers)

LOVE SONG

Across the sky
a lonely requiem of clouds—
The wind's folded hands praying
in the dry twigs—

Moles
emerging from subterranean fortresses—
Snakes as yet
reluctant—

Spring
promising the girls
love's murder
behind garden fences—

At nightfall the pond
offered you
my true
likeness—

(Ewald Osers)

SENTENCE

Whenever sentence is passed
a great darkness falls
on the courtroom
as during crucifixion.

But now, in mid-October,
light's candle
burns with bright flame
about your face.

The dried-up palette
of distant gardens,
false witness against the painter
who has gone away for a moment, no one knows where.

Before I realized it
judgment was passed
on love:
sentenced for life.

(Ewald Osers)

ROYAL PROGRESS

At noon in the garden on a throne of wood
of the kind that's used for businesslike coffins.
Ceaselessly trotting towards nothingness.

In the late afternoon alone on an empty bridge.
The browns and yellows, unobserved, have entered
the foliage of trees on the island below.

At nightfall in the ermine of darkness
along the royal road up to the castle
captured in the last battle of all.

Under the clay's surface the dampness of rain,
street lamps retreating towards the past.
In the night's gateway a dustbin with scraps of voices.

At dawn we trip over our own shadow,
unknown posthumous sons
of nameless kings.

(Ewald Osers)

MIROSLAV HOLUB (b. 1923)

Miroslav Holub, born at Plzeň (Pilsen), is a distinguished research immunologist as well as a poet. Holub spent two years (1966 and 1967) doing research in the United States, and in 1979 he was Guest-Writer-in-Residence at Oberlin College. The volume *Sagittal Section,* in translations by Stuart Friebert and Dana Hábová, published in 1980 in the Field Translation Series, is a direct outgrowth of this engagement. Dr. Holub is the author of numerous books of poetry, including *Denni služba* (Everyday Duty, 1958), *Achilles a želva* (Achilles and the Tortoise, 1960), *Slabikář* (Spelling Book, 1961), and *Tak zvané srdce* (The So-Called Heart, 1963). He has also published prose works on literary subjects, and is the recipient of numerous awards for his contributions to literature.

EXPERIMENTAL ANIMALS

It goes easier with rabbits than with dogs or cats. An
experimental animal should not be too intelligent. It is
uncomfortable when its actions resemble those of humans,
it is uncomfortable when you can understand its terror
and sadness.

But the saddest thing is to work with newborn pigs.
They are ugly.

They don't possess or want anything beside their spring of milk.

Their hard and awkward legs buckle underneath them, their
snouts and tiny hoofs are extraordinarily useless.

They are ugly and stupid.

When I have to kill a piglet I always pause for a while.
About five or six seconds.

About five or six seconds in the name of all the beauty
and sadness of the world.

— Get on with it, — someone says then.

Or I say it myself.

(Daniel Simko)

THE FRANKLIN BRIDGE IN PHILADELPHIA

Steel blue clouds and the enormous brown Delaware river.
The bridge like a prehistoric metal dragon. The parabolic
wings of hanging pipes, floating ribs, asphalt skin of the
six lane highway.
The endless drone of automobile herds.
The sky, river, and bridge.
Wind and engines.
A new countryside. I think I like it.

(Daniel Simko)

THE GILA DESERT IN ARIZONA

We are travelling for the sixth hour, rocks, sand, idol
serving cactuses, the bell of night shivering from one end
of the horizon to the other. The sky, cut by the rotating
disk of this planet, bleeds bluntly.
Joseph Conrad wrote:
I remember my youth
and a feeling which will never return,
a feeling that I can live forever,

220

longer than the sea,
or the earth,
longer than all the people.

(Daniel Simko)

A ONE-WAY CONVERSATION ABOUT RADIO CITY MUSIC HALL

— But the best is Radio City Music Hall, one respectful
and simple man told me. You have to go there. We go there
with my mother, and she's already eighty. It has 6,200
seats and on Easter there is a ceremonial performance
with rabbits eight meters tall.
— And simultaneously 36 ballerinas in special suits
dance there, one just like the other. And those colors
all around, just like a rainbow. And the biggest orchestra
in town.
— Imagine, one just like the other, and all of them are
beautiful. And all of them do everything exactly at the same
time. Isn't that art, what with all those girls and all those
colors and all those decorations and all that music, no?
No, can sometimes be said very quietly.

(Daniel Simko)

BRIEF REFLECTION ON CHARLEMAGNE

Outside in front of the gate,
a bell hangs. Charlemagne, son of Pepin the Short,
had it put there. Those who've suffered injustice
can ring and Charlemagne interrupts being king,
receives them right away, listens to them
and metes out justice.

This happened in Eight Hundred.

The bell rang this year.
In the rain, more like
collapsing than dripping,
and lasting eleven hundred years,
drenched, seedy, sheepish,
in a clown's costume,
stood Charlemagne.

In broken Frankish, he demanded
a hearing.

(Stuart Friebert and Dana Hábová)

CROCHETING

With hooks delicate as the arms of stars
she twists the days and nights together
into an endless pullover.

She's the one
who'd dress the rocks in chenille,
draw the nautical miles of a ship at sea
through the soft tunnel of a sleeve
and wrap a stratospheric shawl
around meteors for warmth.

And yet
we walk around naked,
naked and cold,
sonny boys.

(Stuart Friebert and Dana Hábová)

THE GARDEN OF OLD PEOPLE

Malignant growth of ivy.
And unkempt grass,
it makes no difference now.
Under the trees, the invasion

of the fruit-bearing Gothic.
Darkness set in, mythological
and toothless.

But Minotaurus beat it
through a hole in the fence.
Somewhere, Icaruses
got stuck in webs.

On a bright early morning
the bushes reveal
the unabashedly gray, impudent
frontal bone of fact.
Gaping without a word.

(Stuart Friebert and Dana Hábová)

JOSEF HANZLÍK (b. 1938)

Born in Neratovice near Prague. After studying psychology he became
poetry editor on the staff of the *Plamen* literary monthly, which
discontinued publication in 1969. Also a translator of American, Russian,
and Yugoslav poetry, he has written for children and has had considerable
influence on young people. He has published numerous collections of
poetry, among them *Potlesk pro Herodesa* (Three Cheers for Herod, 1967),
from which "The Cottage behind the Railway" and "Who's That Driving a
Black Cart..." are taken.

THE COTTAGE BEHIND THE RAILWAY

Yes I too have a cottage with a moss-grown roof
which moults like a horse one pities too much to shoot
And whenever a gale blows down the wires which
    stretch between us
and from light to light a lunatic candle flies under the
    ceiling

beating like me its head against four walls Then I hear
the sounds of the night trains—the voices of rails and
    wheels and the groan of the sleepers

and sometimes a call—like a cry answering my anguish—
and if something suddenly tinkles against my window

then surely it is an engagement ring or the key
to the town beyond the hills I throw my window open
   to the dark
the rain strikes my face and blinded by it

I clutch its threads like a great fair mane
which waves from the train its hooves pounding
   escaping
past all returning—like the horse of our century

(Ewald Osers)

WHO'S THAT DRIVING A BLACK CART...

Who's that driving a black cart
through the disconsolate rain
who's that not sparing the exhausted horses
wheels drowning in the mud

Who's that driving a cart
through this landscape with no roadside inns
and no burning pinewood
disregarding the dawnless night

Who's that driving
along this futile road through a drowned world
emaciated to the bone

Who's that driving an empty cart
who but the gravedigger who like the captain
dies last

(Ewald Osers)

VARIATION VI
(Question)

Where are you going? Where aren't you going?
And will you get back dead and well?

Are you as white as the milk of God?
Does Aunt Katherine drink beetle wine?
Did we have one foot in paradise?

Signor d'Amici
what was that story about the Lombard sentry?
Shall I too be shot through the heart?
In what flag will you wrap me?

Did you lock up when you left the house?
Did you switch off the light?
So no one else should come
and make himself at home?

Did I ever see anything
in any country anywhere?
Will a dream like that bring luck?

Shall we change water into poisoned wine?
Will the air kill us as it strikes us?
Shall we fall under the wheels of a car tonight?
Can death do its job even without glasses? Can it do it by heart?

Can you hide in yourself
so that you can teach him to sing
even before birth?

Can I hang every other child by the dangling rope?

Is truth in lying stockings
a true lie?
Can one overtake the past
and return to the future?

Will I die?

(Ewald Osers)

225

ANTONÍN BROUSEK (b. 1941)

Born in Prague in 1941, Brousek absolved his studies in Czech, Russian, and philosophy. He has worked in the construction industry, with Prague Radio, and in publishing, where he is employed at present. *Spodný vody* (Undercurrents, 1963) and *Netrpělivost* (Impatience, 1966) are his first two volumes of verse.

From PILLARS OF SALT

1.

The walls shriek
but they don't hear
the pavements heave,
they don't see, they're fleeing
through the smoke
which lifts them up,
lighter than air.

From clouds oppressing
the ground the smack
of the first bullets,
blisters of asphalt
in the puddles
licked by the flames,
earth's gooseflesh:
a grave

each backward glance
at the city behind them
which is slowly collapsing.

2.

Where are they now,

caught up at last with those before them,
dropped breathless among their own kin,
passing on their outcry,

hard to say
now
where they are.

Scrubbed out their names,
expunged the inscriptions,
the last traces leading to them,
and the late flowers
turned sear by frost and the hiss of
spitting into one's own face.

Perhaps they know by now
and have turned
in their graves, face into the clay,
sobbing, sobbing,
shaking the earth
on their backs,
from the eye's deafness jerking
a pinch of salt.

7.

I would have left long ago
had I not been detained by words
which it hurts me to give up,
cumbersome luggage
above the permitted weight,
so heavy that I know:
I shan't get far with it.

And so my heavy load
slips from my hands,
the further I go the more often
I stop
to recover
my breath;
each stop
a backward glance:
how far have I come
and how far yet.

In my rucksack an egg
which will not hatch
and a mineral grown from poems,

salt:
not of high quality
but good enough
to sprinkle on one's bread.

I go,
step by step,
scattering salt
for the way back.

8.

And what is it really like
where so many have gone,
where your eyes can never reach,
for the ground opens
and you look back.

A stalactite cave
spattered with clay
through which grows
a tangle of minerals;
a colonnade of salt pillars
hiding the spring
of dead water;
in the gravel a conch
shaped like a pelvic bone,
and in it, encapsuled,
the sea
and a tear,
akin in the sign of
salt.

(Ewald Osers)

MIROSLAV VÁLEK (1927–1991)

One of the most distinguished Slovak writers, Válek prepared for a career
in economics and has been a recipient of some top public appointments,
including the posts of Vice-Chairman of the Czechoslovak Writers' Union
(in 1968) and of Slovak Minister of Culture (in 1969). He has also been
active as an editor of a variety of literary periodicals. Among his collections

of verse are *Prìtažlivosť* (Attraction, 1961), *Nepokoj* (Unrest, 1963), and
Milovanie v husej koži (Love in Goose Flesh, 1965).

THE APPLE

The apple from the cupboard rolled to the floor.
Pack up your things and go.

She leaned back against the door
and with her eyes screamed:
For god's sake, please, no!
But I knew at once that I had had enough;
I got to my feet,
picked up the apple,
dusty and still green,
and put it on the table.
Incessantly she begged, she came to the table,
and cried.
She looked at me, wiped the apple,
and cried.
Until I said: Put that apple down and go!

The events unrolled as I had envisaged.
What does it matter if the sequence was different!
She opened the door,
I turned pale and said: Stay!
But she packed her things and went.

The apple from the cupboard rolled to the floor.

(Ewald Osers)

STEEL-ROD BENDERS

On Vuk Karadzic Street
on the corner
two men are bending steel rods.
It is, in a way, poetic work.
The iron, like a verse, resists the human will.
But verse after verse,

verse after verse,
and you have a unique poem.
Above all those verses,
on the third floor
a blue-eyed punch-line settles in.
It will lovingly feel its four walls:
How firmly constructed it all is
and how much strength must go into such a wall.

If I'd had bigger muscles I'd have become a steel-rod bender.

(Ewald Osers)

HEARING

Terrified
he jumps out of bed,
he hears the roar of distant galaxies,
the remarkably high-pitched whistle of falling stars,

he realizes
time and again
his abnormal sense of hearing.

Hands clapped to his ears he stands in the middle of the room,
louder and clearer rings in them
the thunderous surf of darkness.
When at last he falls asleep
the birds awake in the lime-trees under his window.

I saw him again.
He was walking on the pavement
like a man enthralled by something that's inside him,
he broke his step,
and at that moment suddenly
heard his feet rubbing against each other,
the crackle of sparks,
the confused trampling of urgent words.
He turned his head,
saw hair that was burning,
felt sweet burning vertigo
and nothing more.

Of course, that's all speculation, said the driver.
I honked like mad,
but he was looking at that redhead,
didn't move at all,
and now he's a goner.

Who'd have believed it,
such a fine-looking fellow
and deaf as a post.

(Ewald Osers)

S. DANIEL SIMKO (b. 1959)

S. Daniel Simko was born in Bratislava, Czechoslovakia, and came to this country shortly after the events in 1968. He is a writer and a translator. His poems have appeared in numerous magazines in the Midwest. He is at present working on a book of his poetry, and is translating Miroslav Holub's selected prose work. Mr. Simko lives and works in New York.

THINKING OF MY FATHER ON A BUS TO BALTIMORE

Before I can even look
the earth outside the window
becomes heavy.
This day
already filled with light
slants from a ditch.
80 miles from there
crows seem knifed against the sky
rolling over the fields.

I suppose your steps are still hard
and make a sound of pigeons landing,
your brow is still creased and perspiring,
your hands in your empty pockets.

How many thoughts circle a very tired head
that you suddenly place in my lap
in a dream of your childhood across the sea.

The night around your shoulders,
you went to sleep and said nothing.

Nothing can be heard under the soft branches
of the tree where your father died
and returned in your dreams.

There are no sounds,
not even of this match
I strike, to see you better.

THREE SONGS
   after three studies by Käthe Kollwitz

1.

Walls are leaning,
gathering something
from the wet earth.

A child strokes the mane of a horse
drinking long and hard after a run to no place.

Rain hits the windows hard—
inside, a woman slaps bread dough on the table.

Two workers exchange hands,
a man dies pruning a tree...

Outside, a man is sleeping so deeply
it looks as though he were about to speak.

2.

A string slapped the hands
between the knuckled fingers.

A clock stopped.

The seamstress fell asleep,
cloth in her hands.
The nakedness of the needle
holding a stitch of thread.

A roof fell silent
over someone
learning to speak
last words.

3.

In the distance the land
heaves with wind,

trees bolt upright
by the river.

Water is bitter to a tongue
ringing with song

about to end in a nearby grove.
The women carry

buckets of water to wash
battered clothes

longing for a body.

COMING HOME

1.

Sunday, Churchbells. You stood outside
in a strange town, weary and cold, in a coat
stolen from the lost and found. Steps on
the pavement. Bags carried carefully as
children. You pulled up the lapels of your
coat and began walking toward the old church
in the center of town. You could hear voices
inside saying something softly in a language
you did not know. Someone inside suddenly spoke
your name and you began walking away, frightened,
toward the first doorway. I could hear you. You
were breathing in my sleep 8 thousand miles away.

2.

You were breathing in my sleep. A moth was beating
itself quiet against the panes. At night you lit
a burned candle hours after the lights went out
and sat at the window singing softly to yourself.
Twenty one years later you come back with
careful steps, no one's, the Ohio rising
out of its banks. A few swallows. And the wind
rocking them. Bells. You are alone.

THE ARRIVAL
after a photograph almost taken in Berlin

Wet slate roofs. Pigeons. A light.
A leaf on the sidewalk.
The shadows slipping between cobblestones.

It is already dusk
when you arrive
from Paris,
smoke rising from the Diesel
as you step out
with your black hair untied.

I am almost always
turning into that smoke,
into the pigeons landing
on the glass roof.

Or I wake up
and you come
with a shawl
black with stars.

Paris, 1980

THE JEWISH CEMETERY IN PRAGUE

It gets harder to walk
out of those crates which stop
light from entering.
The sky, bending over a body
touches the pale skin of grass
like the fevered head of a child.

You've left behind the town.
Trees long to lift you
gently from the splintered bed
into the wind. There is none.
It has stepped out of your bodies.

HUNGARIAN

INTRODUCTION TO HUNGARIAN POETRY

Hungarian poetry of the postwar decades has remained true to its image as the leading genre in its republic of letters; for better or for worse, Hungarians remain a nation of poets. To be sure, most of them have worked in at least one additional genre, and it would have been good to open our roster with that multigeneric artist par excellence, Lajos Kassák (1887-1967), avantgardist poet, novelist, playwright, critic, painter. Apart from the issue of genres, we also had to stop somewhere, certainly just short of the generation now about forty years of age or younger. For fine examples of work by three men in this latter category, Dezsö Tandori (b. 1938), György Petri (b. 1943), and Lajos Pintér (b. 1953), see "Three Young Hungarian Poets," introduced by Kenneth McRobbie, in the Summer 1981 issue of *The Kenyon Review* (Vol. 3, no. 3, pp. 95-104).

Twelve poets, born through the mid-thirties, most of them multigeneric, take their places in the present section, and at times it may seem tempting to feel that the amount of time a poet spends writing and publishing poetry rather than, say, fiction, may ultimately help determine his quality as a poet (as distinct from such ambiguous matters as reputation) to such a degree that we can distinguish even on the basis of the smallest of selections. This may seem true, in our instance, of writers born before World War I. Sándor Weöres, for example, many feel to be a far more intensive poet than Gyula Illyés, although Illyés's lyric output is enormous. The older poet is a distinguished multigenre writer, sociographer, biographer, journal editor; the younger is an anthropologist and linguist among poets. The approximate converse may also hold. Judit Tóth and János Pilinszky are equally fine (although very different) poets, but their achievements in other areas are not equivalent. It is almost impossible to judge "purity" of achievement in poetry on grounds of generic interests alone. How a poet's "other career"—as a playwright, memoirist, scholar—affects the poetry that should be his primary concern, deserves to be judged on merits of the individual case.

What answers have Hungarian poets found to the problem of spiritual and intellectual moonlighting? One of the fascinating, and certainly most convincing, answers seems to be that working in other media strengthens the poet rather than weakens him. Even in Illyés's work we feel an intimate connection between the poetic self and that other. In "Part of a Novel" we do not feel far from the writer who has so successfully shown the people of the plains how to overcome poverty and enter upon the opportunities of a civilized existence. "Spacious Winter" seems to hold at least the suggestion of an answer. Going on to István Vas's voice of meditative wonder, as in "What Is Left?" or in "Catacombs," one feels that the careful, guiding eye is also that of the prolific essayist and autobiographical novelist. At least of Vas it is safe to say that he consciously defines himself in all the genres in which he has worked; from the poetry there emerges one of the most consistent urban sensibilities of all of modern Hungarian literature. Not many have written about Budapest the way the poets of Prague have of their city, but if anyone in our day has, it would be Vas. Among the younger generation Ottó Orbán shows some of this capacity; in prose poems and, among the poems below, in "Snowfall in Boston" he has shown a love for cityscapes especially as a means for depicting human frailty, as well as a sense of man's falling victim to the environment he once created and of which he has long since lost immediate control.

A strange realm, the world of nine to five—of noise, pollution, traffic snares, and worn nerves. Sándor Weöres and Ágnes Gergely both combine their vision of the city with the mythical and the exotic. In Weöres's poetry individual portraits and scenes take us to such places as China, or else to an envisioned "palace" of the related baroque splendors of Shakespeare and of Velázquez; in an epic poem Weöres offers an argument according to which human consciousness antedates our planet, and our present tellurian plight is the result of space travel following our expulsion from a bubble-shaped interstellar paradise. It seems clear that Weöres's own intratellurian travels have many times taken him between the realms of poetry and verse translation, from the imagined Chinese temple to the poets of the real one. This is also true of Gergely, a voluminous translator of works by English and American writers, also of another woman poet falling between Weöres and Gergely on the chronological scale, Ágnes Nemes Nagy. While Nemes Nagy has brought close to Hungarian theatergoing audiences the baroque of Racine and of Brecht, she herself, in her precisely crafted verse, has created a myth of Egyptian pharaohs, trees, real and mythical beings. Her "The Horses and the Angels" holds its peculiar charm in the fact that, unlike Rilke's angels, hers are tangible, almost human; their behavior overlaps in a bemusing way with the hapless and absentminded city dweller's. No wonder the poet wishes she could hang onto them as to alternative presences.

Among the works of the younger poets, certainly the most provocative achievement has been Ágnes Gergely's multigenre poetic tour de force, *Cobaltiland*. Realm of a whimsical cobalt-colored king, Encelado Sulfato XVII (whose name, discovered by Weöres, belonged to some obscure Hellenistic king of nether Asia), the land of the Cobalts is disturbingly like and unlike the "kingdom" of our everyday preoccupations. Periods of history are, to be sure, telescoped; the king can act like a child; government officials can be like paper cutouts in some shadow theatrics. But the inner realm is that of the realities of Western culture, and includes such unexpected events as a poem in memory of the late W.H. Auden, or a piece of psychological counseling ("Gnosis") offered to a monarch who, throughout the book, badly needs it. In "Encelado and the Facts" the hapless king is confronted with nothing less frightening than the public misuse of language, the euphemism and the threadbare aphorism, such stuff as the nightmare of commercials is made on.

Let the reader be tempted to go on commenting on his own favorites, among which, it is to be hoped, will be Márton Kalász's "Maiden," Sándor Csoóri's "The Tenant Who Lives in Fire," Görgey's "Thomas." Without being concerned too much about "genres." Let the poem be well written and then recognized. Ultimately, little else matters.

GYULA ILLYÉS (1902–1983)

Illyés is regarded as an important transition figure and the ranking living Hungarian poet. The son of a machinist employed on one of the large landed estates, Illyés completed his schooling by efforts on the part of his entire family, and went on to become one of the important populist writers. His prose classic is the sociographic study *Puszták népe* (People of the Puszta, 1936); his biography of Petőfi is also regarded as basic. Illyés became an editor of two of the most important literary periodicals, *Nyugat* and its successor, *Magyar Csillag* (Hungarian Star), the latter suspended with the German occupation of Hungary in the spring of 1944. Since 1928 Illyés has published some twenty-four volumes of his poetry, including several collected and selected editions. His works, in several genres, have been translated into all the major European languages.

SPACIOUS WINTER

Our respective views did better, finding each other
in that spaciously echoing and excessively lit
colonnade which that winter morning made.
Sunlight, with the sensitive touch of a doctor, walked
on the fragile, small branches of glasslike trees:
immediately they reacted by shaking down frost.

239

Everything around us was more finely shaded, more capable
of detailed rendition—and inside us—brought on, too, by the fact
that from under your sole and from under mine, the snow
crunched—conjured up that squeaking of childhood-new
shoes, and with it, the Eden-age of guilelessness.
With the courage of moving into a
new apartment, refreshed by the smell
of still-drying whitewash, we walked
illumining-clarifying all: how we should
refurbish with miniature furniture
—for it *is* possible!—our lives.

(Emery George)

SOUR-CHERRY TREES

As is a face, the throat crushed,
so is a cherry tree: deeply flushed.
That is the shade of red it turns;
the entire tree now chokes.

I can picture just the throat
which would tolerate such a threat,
and the strangler with the gall
to stalk these vineyards in the fall.

I come from below, stop here, on further,
look upon the scene of murder,
walk right on into the hill,
like one free to flee the kill.

Near, far, on these slopes, there are more
crimson trees than ever before.
Nightmare—I find strange consolation,
beauty in the very notion.

(Emery George)

PART OF A NOVEL

In the courtyard a poor blind man is repairing a concertina,
ping, ping... he tames his destiny's dark angels with music.

He presses his ear to his instrument, smiles as he tries a tone,
then, face to the sky, he starts up a song, taps with his tiny feet.

He smiles and taps; he probably thinks he's flying.
Three smutty children admire all this around the blind man.

Triumphantly he flies to Jesus, forgetting he's been drunk since morning!
Into his whisky bottle the three little hoodlums are littering.

The blind man sniffs, spits, yells, and flails about,
in his darkness he's fighting with monsters of the underworld.

Four poor women run out into the courtyard, and behind the children
they weep and bawl in a hubbub, like female devils.

At the end of the courtyard, coming from the West Depot, a freight train
                                                                    rumbles by
it buries their awful world in smoke and thundering noise.

I'm looking for my aunt here, the blind zealot; as if from under earth,
I hear her squealings from the hell of penniless indigence.

(Emery George)

DEEP DUSK

Deceit goes well with a Fall afternoon.
The sky is sparing with its light
but it strengthens us to face loneliness.
The remains of the bright blue of summer
scatter from above the houses
like the parade of villagers
coming from the marketplace.
The smell of flowers is chilling.
The plowed gardens smell sour
like freshly-dug graves.

241

The man sitting in a chair
would feel warmer with a coat on.
Dusk pulled a book from his hands.
The theater in the sky keeps us bewitched;
it's heightened with falseness,
laced with a dark prophecy
that gleams and becomes true.
Two hands seek shelter
in each other's warmth.
They're knowing and prepared.
Instead of pears
trees let fall sad gestures—
the waving of a hand.

(Nicholas Kolumban)

ANNA HAJNAL (1907-1977)

Poet, translator, author of books for young people, Hajnal is a seminal
figure in contemporary Hungarian poetry, and one of the most
distinguished women poets of our time. Since 1935 she had been the author
of fourteen volumes of verse, including *Ének a síkságon. Összegyüjtött
versek* (Song on the Plains: Collected Poems, 1977), published shortly
before her death. Books by Anna Hajnal have been translated into German
and into Chinese, and she has been prominent in a number of international
poetry anthologies.

THE JAGUAR IS GETTING READY

Everyone will have his turn:
until then move safely, slowly,
taking cover, in shadows, in tall grass.
Just look and perceive wisely.
Just be, mutely, simply
and hope, as we all hope
that: today it is not yet my turn,
not for me is Death swinging
on experienced, tranquil branches.
Today it jumps, not on me,
                    but I know

that in the end it is always the same:
I am a turtle and the taloned jaguar
will flip me over. Then he will scoop me out.

(Juliette Victor-Rood)

AFTER ALL

After all, what have I become?
The island Iceland in a blind fog.
Gliding in the far north.
I swim in mushy ice-water.
An ice-barrier surrounds me,
to protect me?
                    protect from what?
What boils in me darkly,
bubbling, swirling upward,
melting my thick cover:
the firmament may blanche
while being sliced upward to its lap
by a foaming, vapor-tressed head
ragingly crying: the geyser.

(Juliette Victor-Rood)

THAT'S ALL?

Shearing, as the gardener
snips the sucker,
controlling wild growth
with shaping hands,
looking and choosing—
which bud's to be the branch—
rooting out, cutting or pardoning
by design and scheme:
trimming pyramids, tall arches,
scissoring bowers for gods—
how I'd love doing that—
taking hold of the passionate growth
in my unmastered heart.

Slicing through wild, winding
trailers, charming
with a bright, sharp blade—
to but loosen its hold on me!

release its hold?
and must the clasper wither?
trailers, leaves, tendrils droop?
A French park, my loving?
moderation, cautious suffering?
precise forms, narrow blossoms,
the reign of geometry,
is my calmness to be a tight calmness?

(Jascha Kessler)

ISTVÁN VAS (b. 1910–1991)

István Vas studied in Vienna, but returned to Hungary in 1929, and has
been publishing his poems in the liberal weeklies and periodicals since then.
An active dramatist, essayist, and novelist in addition to being a prolific
poet, he was, especially during the twenties and the thirties, a translator of
drama and of prose fiction from the English, French, and German. His
translations of poets are collected in the volume *Hét tenger éneke* (Song of
the Seven Seas, 1955). Between 1932 and 1976 Vas published over twenty
collections of his own verse. A three-volume collected edition of his poetry
appeared in 1977.

WHAT IS LEFT?
Jeu de Paume

And what is left for the others?
The objects, seas, the gardens,
Enormous magic of light—Monet!
The moment found, new sesame, the great fever,
Melting into flames, Rouen Cathedral,
As transformed by dawn, by night, and by day.

Then the races, circuses, cabarets,
Regattas, nightclubs, that mix of *modern* ways
Their strong eyes braided into a worldly halo;
But he didn't live among feasts, no, the light
Of glory he found was not of our sight,
Around things ancient: Vince, the holy fool.

The chairs, the sunflowers, Auvers' wild temple,
The fibrous laying-on, the rough texture,
Soon—you feel—strong, attacking vegetation,
Trunks, leaves, clods, grains of sand,
The host of plant life roused to dance,
Will swallow the whole composition.

He himself: just like an insane sunflower.
Do you recall the self-portrait next door—
Old Tintoretto, bidding us farewell?
White beard, intelligent, tear-ready eyes;
But he'll not mourn what couldn't be; he bewails
Only what's his, and what he stands to forfeit still.

But Van Gogh suffers, vibrates here on the canvas,
And the blue eyes stare at you defenseless,
Thickly the blue and yellow brush strokes swirl;
Tormented, frightened sadness gone astray,
Out of his beard martyr-bright roses spray,
From his auburn hair too you see madness unfurl.

And brave of the brave, the greatest, oh I love you:
Lautrec, lord, beggar, priest, and merry-andrew,
Cruel sufferer, brother of Baudelaire!
You who feared no demons of flesh, of absinthe,
On your canvases we see the whinnying dance
Of Paris's happy ghost, his skeletal stare.

And magician of movement, Degas, who observed
The defenseless body, distorted and sad,
Though in the footlights and through veils of tulle;
You, who illumined poor little monkey-faces
Of ballerinas in motion, their faerie-graces
Gleaming in a milkwhite, goblin-blue circle.

And the great flight! Primal dreams, craving-racked,
Tahiti! Many-colored faith, that once turned fact.
Epidemic spreads, a learned superstition's hold.
Animal angels, Eden far off, naked;
In the picture's corner some writing, opiated,
And their bodies' gold... their bodies' gold...

(Emery George)

CATACOMBS

Today a silver coffin guards St. Agnes'
Ashes; above it stands a statue of bronze.
But what a narrow bed, in other ages,
Poor cellar-soil, held her in their bonds.

Clumsy drawings, bones, a deepening—
The crucifix just barely has its place.
Was then this desolation everything?
It's the birth of all that shall be greatest.

And now the Age of Constantine, right soon....
Later the pyres, bells booming with lust;
The gold-benumbed frown of Byzantium
Answered the martyrs fallen here to dust.

To embrace the stone blocks of dank cellars
While, outside, authority's storm howls—
Those who survived themselves turned killers,
Received their share of eternal power.

Wearisome circus! Iron bylaw! Away!
Ancient aqueduct to where I'm reborn.
All catacombs answer my martyred brain—
Grotesque experience, embittered form.

(Emery George)

TAMBOUR

Whence, hardworn drum
winging past death
natural, unnatural
through siege and shelter

Wan goatskin face
its spots of coalcellar dust
raised by dancing, ingrained
by the sweat of palms beating

Fingers dancing its sticks
glossy wood, bloodclot-brown
flat palm flicking
counterpoint, tapping soft
syncopated commands staccato
on the smooth parchment

Drum, resurrection drum
out of its opened mouth
swarm the raised knee
taut rounded buttocks
tight legs
leaping burning soles
the hair floating in flight

Sounding box, drumming
its wan death's skin—
victorious, true portrait
its blackened face
making its accusation
with the searching, sceptical hand
the fugue erupting, shaking
its captive soul, the dance

(Jascha Kessler)

247

Their auburn or black hair, curly hair
or straight, their eyes, dark, brown, gold
or green eyes, their noses, straight, hooked,
long or pert, retrousse, noses, their mouths,
thick, rippling, thin or hardset mouths, their hands,
white, childlike or whittled bones, active hands,
stained by paint, sun, nicotine,
their magical fingers, their breasts, just barely rounded,
small hills or divine, muscular breasts,
their feet, tender and small, or strong, sleek feet,
their shin-thews, ankle-power, their yellowed, stamping,
pink, earth-strolling soles, long thighs, the flexing arches
of buttocks we'd have mentioned, had we been brave
enough, though for that matter Attila József discussed
a girl's stomach, the slag revitalized in the gut
tunnels and hot kidney pumps—but one thing
we've never said before, not even its name,
except in execration, the name of our one joy,
and of the shells', the calyxes', of wings, altars,
winged altars, of the incendiary, maddening,
silken vehicle of flavor, spice—this fourletter word
we've never mentioned, never this name
we'll know longest of all.
　　　　　A splendid death I'd have,
beyond pain, consciousness, beyond the edges
of the uttered word, if I said no more from
these compressed lips of mine, no more than this,
nothing but this, and only this.

(Jascha Kessler)

SÁNDOR WEÖRES (1913–1989)

One of the most skilful, versatile, and language-conscious of contemporary
writers, Weöres was born at Szombathely, in western Hungary, and took
his doctorate in philosophy at the University of Pécs. Of precocious poetic
capacity, Weöres was recognized as holding out serious promise at age
fourteen. He was twice awarded the Baumgarten Prize (in 1935 and 1936),
and has been devoting full time to his writing since 1951. Widely traveled,
he visited the Far East in the thirties, returned to China in 1959, and has

twice visited the United States since the sixties. Weöres is the author of many volumes of his own poetry as well as of translations from a great number of languages, including those of Africa, the Orient, and Polynesia. He has also shown an interest in re-creating the poetic language of earlier centuries, as in his recent *Psyché* (1972). A monumental three-volume edition of Weöres's poetry and other writings has just appeared.

TO THE SPIRITS OF SHAKESPEARE AND VELÁZQUEZ

I turn to you, in the deep twilight of a cellar,
letting out sprouts, I dream of light, of sky, of space,
confessing I don't read your language, Insular;
you Other!—of your brush I haven't seen a trace.

Across the village meadows I trail my heavy shade,
dishevelled, muttering, like one who has lost his mind,
so I pull my progressing poems into shape,
desire lifts me toward you—I stumble on a clod.

A tired old canal, my little town's horizon,
and mumbling jailers both: a river and a mountain,
and molded hills surround my home town's gentle nest;

as fish by traps, I'm guarded by thickly-woven malice.
O Shakespeare, O Velázquez! inside your gleaming palace,
on marble stairs I sit, poor, grimy, and oppressed.

(Emery George)

ECCE HOMO

Eye: solitude.
Depth: road, height: mist;
across: mountain carries you.

Skull: cell.
Outside: locked,
within: infinite.

Open eye, closed in fog.
closed eye,
open within.

Body: circumscribed form;
an eagle flies
above itself.

Torchless star
in dark;
rises in darkness,
King of Light.

(Emery George)

CHINESE TEMPLE

Sage	high,	Four	then
grove,	low,	bronze	deep
dark	night,	gongs:	peace,
bough:	king:	Life,	tones
green	come,	rank,	like
wing	blue	fame,	chilled
spread,	shade.	long,	song.

(Emery George)

ELF

"My eyes are red, my mouth black;
haven't seen you in three weeks...."

On the dresser's shadow he stands,
transparent as a drinking glass.

"Come here, shadow-operator,
let me catch you with the tweezers."

Gives the furniture a crack;
vanished—never once looked back.

(Emery George)

SATURN SINKING
To the memory of T.S. Eliot

They took my flock—I could care less! No more work for me,
no worries; an old man has it easy in a nursing home.
First they trounced the priest, horned, delirious speaker,
where, from his desk, he'd fly to heaven daily—what a fool!—
and chose smarter priests; later the king, unarmed protector,
and hired kings with swords; then the philosopher (for,
after all, we have plenty of *them*); finally the poet
(what does he count his fingers for, mumbling?); in his place
come purposeful songwriters, to taste, on commission.

So I stand, face to the wall, with my broken shepherd staff.
My herd swarms the trough (oh, for all the cheerful, brand-new
attractions floating around in there!); nose-to-nose, nose doing
nose dirt there—I don't care; it's no longer my trade.
They'd stab me with a tusk if I saw: what's become of huge
population booms, of the scratched-out womb, of greedy
inebriation, of ever-speeding rates of velocity, of murderous rays,
of the bomb planted in the doorway—

                    It's like the train running
on its tracks toward the precipice without an opposite shore
—what do I care!—; could be, they'll stop it the very last minute;
else the tracks extend over the chasm; I'm just blind;
could be: on the drop's edge the train spreads wings, flies up—
they're supposed to know it, not I. Too bad if they don't.
What do I care at this point? my staff is broken; it's easy
to rest on the straw, to rest up from the pain of
millennia. They can't see me; their heads are in the trough;
I too see only their rumps and flapping ears.

(Emery George)

JÁNOS PILINSZKY (1921-1981)

Regarded as perhaps the most distinguished Hungarian poet of our time; his sudden death late in May of 1981 left the world of East European poetry much the poorer. A Catholic poet of a decided existentialist and agnostic stamp, Pilinszky founded no school; his spare work often leans in the direction of minimalist and hermetic verse. His wartime experiences, most of all internment in several German concentration camps, shape a significant segment of his vision. His principal books of verse are *Nagyvárosi ikonok* (Metropolitan Icons, 1971), *Szálkák* (Splinters, 1972), *Végkifejlet* (Dénouement, 1974), and *Kráter* (Crater, 1976). Pilinszky is also the author of several plays and of important essays in poetics. Selections of his poetry have appeared in England, in the United States, and in West Germany.

PARAPHRASE

For everyone's nourishment
—as it's now on record—
I give myself to the world
as living food.

For all that lives hungers
after the living alone,
be it your best lover, in the end
she'll cover you with gore.

So I toss on my bed
and tremble at the thought
of just who it is I'm feeding
my heartbeat!

What sort of trough is this bed,
yes, what kind of feeder?
and what gleaming, clean longing
thrusts me there?

Heart ceaselessly arriving
for the crowd to chomp it,
I'm living nourishment,
stammering, stomping.

I am your striving sustenance,
without pause, the whole;
let me comprehend your hunger—
digest my soul.

For whoever in the end is no one's
is everyone's morsel.
Destroy me then, terrible love.
Kill me. Don't leave me to myself.

(Emery George)

CELEBRATION OF NADIR

In the bloodstained warmth of the barns
who dare read?
And who dare,
in the splinter-meadow of the setting sun,
at a time of sky-tide
and earth-ebb,
set out on a journey, anywhere?

Who dare
stop with eyes closed
at that nadir,
where there's always
one last wave of the hand left,
a roof,
beautiful face or, for that matter,
a single hand, head-nod, gesture?

Who can
smoothly slip with tranquil heart
into dream that surfbreaks beyond
childhood's sorrows and lifts the sea,
like a handful of water, to its face?

(Emery George)

FRAGMENT FROM THE GOLDEN AGE

Joy precedes it, sudden joy,
that modest, beautiful anarchy!
The land is open, smooth even when disturbed;
you look out on the windscrubbed, barricaded roofs,
on the ocean of roofs and stone:
the twilight wilderness gleams.
What is, is unspeakably good.
From every roof you can see the sun.

Chaotic hubbub opens wide,
on the houses, the houses' firewalls,
in the empty doghouse at world's end
summer is the same, of the age of gold!
And it's that same, same pulsing joy,
it throbs, it beats in the sizzling void;
my heart throws me, tosses me away,
and squeezes me to itself, out of its mind!

What emerges here, from the gleaming sea?
Even when I close them, it burns my eyes;
what outside is white-hot—inside the pupil
is where it's incandescent: here, inside!
It's only with *it* that the world turns bright,
of joy, which is always shy of a name.
As at a gallows, so blinding it is,
and so sweet. It's how all things are made.

(Emery George)

AFTERWORD
For Pierre Emmanuel

Do you still remember? On the faces.
Do you still remember? The empty ditch.
Do you still remember? It's trickling down.
Do you still remember? I'm standing in the sun.

You're reading the *Paris-Journal.*
Since then it's been winter, winter night.
You're setting the table near me.
Making the bed in the moonlight.

Without breath you're undressing
in the night of the deserted house.
You let down your shirt, your clothes.
Your back is a bare tombstone.

A portrait of unhappy power.
Is someone there?
              Sleepless dream:
there's no answer—I cross rooms
lying in the mirrors' depths.

Is this, then, my face, this face?
Light, silence, judgment shatter
like glass, as my face, this stone
flies at me from the snowwhite mirror!

And the horsemen! The horsemen!
Dusk hurts, the lamp injures me.
A thin ray of water dribbles
on the motionless porcelain.

I rattle on closed doors.
Your dark room—just like a mine shaft.
On the walls cold flutters.
I smear my weeping on the wall.

Help me, snow-covered roofs!
It's night. Let what's orphaned, gleam
before the day of nothingness
should appear. Gleam—in vain!

I lean my head against the wall.
From everywhere it reaches to me
mercy's handful of snow—
to the dead, a dead city.

I loved you! A shout, a sigh,
a fugitive cloud on the run.
And the horsemen, in falling, thick hoofpatter,
arrive in the rainbeaten dawn.

(Emery George)

255

IN MEMORIAM F.M. DOSTOEVSKY

Bow down. (Prostrates himself.)
Stand up. (Gets up.)
Remove your shirt and trousers.
(Takes them off.)
Look me in the eye.
(Turns away. Looks into the eyes.)
Get dressed.
(Puts his clothes on.)

(Jascha Kessler)

STAVROGIN'S FAREWELL

"I'm bored. My cape, please.
Before you commit anything,
consider the rose garden,
a single rose bush rather,
or one rose, gentlemen."

(Jascha Kessler)

STAVROGIN'S RETURN

"You have not considered the rose garden,
and you've committed what is forbidden.

"From now on you shall be persecuted
and solitary, like the butterfly hunter.
Get under the glass, all of you.

"Under the glass, pinned by the point of the needle,
shining, a bivouac of butterflies shining.
You are shining, gentlemen.

"I'm frightened. My cape, please."

(Jascha Kessler)

ÁGNES NEMES NAGY (1922–1991)

Ágnes Nemes Nagy has worked on the staff of an important pedagogical journal, and has taught at a *gymnasium* in Budapest. A spare and strict crafter of her poems, she has laid great stress on low volume of output with high quality. For her first volume of poems, *Kettős világban* (In a Double World, 1946) she was awarded the Baumgarten Prize. Three additional collections are represented in her 1969 volume, *A lovak és az angyalok. Válogatott versek* (The Horses and the Angels: Selected Poems); her collected poems, under the title *Között* (Between), appeared in 1981. Nemes Nagy was recently a guest of the International Writing Program at the University of Iowa; as a result we have Bruce Berlind's volume of translations of her *Selected Poems,* the source of a part of our selection below. Nemes Nagy is also an eminent translator of Racine and Brecht, and her translations have been performed on the Hungarian stage.

THE HORSES AND THE ANGELS

Coming

Welcome, beautiful angels.
Have some cellar-apples.
Who sent you for comfort, I wonder?
Here's a Jonathan, here's a Winesap.
I feared that what with the lintel
you'd give your forehead a smack,
and how could I nurse a broken angel?
But you made it without mishap!
Like a flame of kerosene,
you trimmed down small and lean.

Going

First to arise was Ariel,
who sings in the angel-choir.
Next was Raphael.
Don't go yet, please.—But I will.
From youthfulness, laziness, bill-
owy-paced slow Gabriel
set off next.—Don't go.—His face
is fuzzy, like pussywillow.
Last to stay was Michael.

Iron-gray hair abristle
like the fathers, he's such an angel.
—I'll hold on to the tip of your cowl,
I know that the apple was mealy,
but that's all there is, that's—stop!
And he left, after all, Michael.

The Journey

Shoe me a horse, blacksmith,
let him take me home at last.
This old nag looks good.

We'll pull up at the pastry-shop,
for the last time I'll peek in,
as if through sunset's window;
see how the pastries glisten,
all the heavenly marzipan—

and everything after is colorless.
And there won't be a soul to see us.
Just the muffled drum of the clattering
horse, like a heart-beat, receding.

And we'll float along slowly,
some water, some trees, the hush
of leafy silence, something
rustling in the underbrush
perhaps, something running away,
a river perhaps in the forest—
but it's vague—I just can't say—
keep that branch off me! I'll duck.
I hug the horse's neck.

The Horses and the Angels

For finally nothing is left,
only the angels and the horses.
Down in the yard, standing merely,
and the angels in my room dawdling;
at times a hundred of them, nearly—
what can one creature do; all alone?
Paw the ground, then stop,
or occasionally flap its wing,
like a bird fanning itself.

Just standing, and that's all,
Just a sight, just a vision,
just leg, just wing—the road, the sky,
remoteness dwells between—

so far from me, so near me.
Maybe they won't leave me.

(Bruce Berlind)

PINETREE

Large, yellow sky. A mountain ridge
weighs on the level field.
On the magnet-earth motionless
dark iron-filings of grass.

A stray pinetree.
Something humming. Cold.
Something humming: in the bark-shredded,
scaly-rooted pinepost's
immense trunk now travels
a paleolithic telegram.

Above a bird, a nameless
bird in the sky—knitted
eyebrows, faceless
behind it now the light fades,
falling eyelids, blind window—
only humming, only the night buzzing
invisible, from black foliage,
charred wrinkled treetops
whose black heart crackles to a purr.

(Bruce Berlind)

FIGTREES

Stunted
Figtrees
Motionless moonshine

Under the trees under the moon
A flock of belled goats presses on
Now even the figs are bells
The seeds jingle in their cells
Just the black-bronze boom
Of the vast sky is quiet

From metal
Pulsates the
Hush

(Bruce Berlind)

TO A POET

My contemporary. He died, not I.
He fell near Tobruk, poor boy.
He was English. Other names, for us,
tell the places where, like ripe nuts,
heads fell and cracked in twos,
those portable radios,
their poise of parts and volume
finer than the Eiffel, lovely spinal column
as it crashed down to the earth.
That's how I think of your youth—
like a dotard who doesn't know
now from fifty years ago,
his heart in twilight, addlepated.

But love is complicated.

(Bruce Berlind)

IKHNATON'S NIGHT

Ascending and descending you take shape
Living Sun
In the dark you die, in light you resurrect
You throb in my heart
      (Ikhnaton's Hymn to the Sun)

When he went down to the square, the pavilions now
were drenched in the lantern lights,
candles erect in the necks of bottles,
a summer All Saint's Day light,
on plank shelves, in dust,
rose crêpe-paper dolls.

Side of hills tattooed,
hearts stippled on olive-green melon
skins, punctured by knitting needles.
Above, the rippling hair of neon lights.
Hot wind. Scrap of straw.
This night was dark.

He went as if trapped
in the disguise of his shape,
he went motionless. An elevated
train streaked above.

      "Clean your face.
      Put it in the cave of your palms,
      the will is striated,
      lead it to water like a bird.
      lead it to water like an animal,
      wash, wash your face,
      every ray of the sun ends in a tiny hand,
      with the hands of the sun your face shall be..."

Night, its leaden drapes
hang between the lights
between the counters
that gleam like ribbon-candy, like caterpillars,
stifling of candles, rush of wind.

"In the ancient garden
in the garden 100,000 things
under the wafer-colored sky
and you have to swallow the other face
and the green flower, the elder tree
on which Judas hanged himself,
and above the star's bit of green
the unmeasurable in the garden,
if ony you could be so minute, my love,
like a god on a wafer."

And now the tanks came.
        Hills of metal waves
the streets ran from them in stone basins,
soft bodies ran between metal and stone
each one lugging a balloon bundle,
the clatter of falling canvas stalls,
splash of bridge rails,
ashes in the distance, the fine rain of glass,
and between the intervals, there is that which blares
above them
that which blares
above the entire planet.

And he hurtled himself over the parapet
together with the others,
headlong they rolled together,
jaggedly, jolting they fell
on top of each other,
like a machine-gun burst,
an avalanche.

It was foggy when he began to see again.
He lay on the shore. Reeds.
Another body was beside him in the mud,
stretched on its back, so stiffly
as if private snow had fallen on it.
He got up from it. In a single motion
he rose like smoke,
beside or from the body.
He was so transparent.
He rose and lay in a single motion.

And he took it, even when he departed
Arcanely, he took the body with him.
Between the long ribbons of fog
he went leading himself
      left hand held by his right.

(Laura Schiff)

STORM

A shirt races in the meadow.
In an equinoctial storm
it escaped from the clothes line
and now dips above the June-green grass
a wounded soldier's bodiless
choreography.

They race, the linens.
Under the lightning's muzzle fire
the last assault of an army of flags
and sheets, they speed, a strange hiss,
a torn sail, a rag,
in the endless green meadow,
falling and rising
their billows mark
a mass grave of linens.

Though motionless, I step out,
I run out of my contour
after, between them,
a runner slightly more transparent
with extended arms
like a half-wit calling back
a dead tree calling back
its flown bird—

Now they fall on their faces.
And with a white-winged sweep
suddenly the whole army soars
soars like a still etching
soars like the body's resurrection
from a seaborn eternity to a pistol crack.

Behind on the meadow remains
only a pleading gesture
and the dark green of grass. A lake.

(Laura Schiff)

GÁBOR GÖRGEY (b. 1929)

Görgey is the author of four volumes of poems, as well as of plays for stage and television. His 1970 collection *Köszönöm, jól* (Fine, Thanks) is perhaps the most characteristic of this poet's love of nonchalant manner and understatement in verse. *Légi folyosó* (Air Corridor, 1977) is also a source of poems printed here. Two sensational poems by Görgey, "Anatomy of a Supper" and "Interview" (in translations by Jascha Kessler), could unfortunately not be included because of their length.

FOLKWAY

Eyes socketing instrumentation
feathers tufting our heads,
shouldering the construction industry,
rattling our colored beads
and striking flint
at freeway rests,
offering rams up
at our dashboard altars

on we go, like the wanderer of the tales,
with a bindle:
bread in it, salt and the first
stone splinter
of the totem image.

(Jascha Kessler)

THOMAS

The most he dared.
Others believed, scared,
escaping that hideous
certainty—
he took on what he could see,
probed the five bloody
inhuman wounds.

And so lived on, unhappily,
faith's contrary exemplar,
till one fine day a martyr's death
was offered him
with wild cheering,
and he thought, might just try it on,
could maybe help, bring it home at last.

(Jascha Kessler)

FRAGMENTS OF TEN POEMS

First Lines

On the ground floor of the sinking apartment house
It can be found in none of Oxford's 13 sections
The universe expanding inch by inch
What, O fog-wreathed Ossian, have I to do here
Fight the fight, Man, wait for the answer
Comes back the croak of this nuncio extraordinary
They're corporeal, all those guests eating out of that platter
In the dungeons of the underground palace of lava
Maybe it's in the flesh that the soul lives
Still and all, one fine morning the king declares me his son

Second Lines

Sipping their wine, they don't even know it's sinking
The absolutely perfect synonym
According to the vague definition of verminkind
The Celts grow mistier in the mists
Happy are they who gave birth to their god on the green hill

Spelling out revelation
And rapping it along to the continents next door
The sinews of Atlas are trembling
Somewhere the ocean is spilled
Twin-tongued flame erupts from the peony

Last Lines, Presumably

Crammed with Riesling, dead men lie stacked in the cellar
So we're better off moving to Cambridge
What a drag, what with all these personal effects
The only thing needed on Naxos was a Shirt of Nessus
Every feeling here is shivering in its sealskin with the cold
O, what it must have been like up in that divine aspic
It's growing like some explosion in the silent flicks
Nothingness dropping away from under the sole
A face helpless in the rubble
Letter and spirit

(Jascha Kessler)

SÁNDOR CSOÓRI (b. 1930)

Born in Zámoly, in Transdanubia (western Hungary), into a peasant family. After a long illness he worked on the staffs of several important literary magazines, both in the vicinity of his home and in the capital. For his poems, of which he has to date published six volumes, he was awarded both the József Attila Prize and the Herder Prize. He is also author of a travel journal recording a trip to Cuba and of a volume of studies in sociography. A novel by Csoóri entitled *Iszapeső* (Mud-Pack Rain) appeared in 1981.

A STRANGER

Someone comes knocking, he wants lodging,
but like an android only capable of grinning at the doorknob,
at the rabbit ears' yellow flowers.
I mumble at him: Sorry, we've just had a death in the family,
we'd like to have a quiet wake,
talk to him beside the bread and wine,

the way our elders did,
and be together with the forest's sound he used to listen to.
He nods, standing there, a stranger in the midst of my mourning,
although he's Earthborn too.

(Jascha Kessler)

MAYBE ELEGIES

Maybe elegies have ruined me,
those slow, hovering, promiscuous birds,
because they taught me that I might live long, long
beneath a sky burdened with my name.

Which is why I've always loved big, lazy waters
that carried trees, suicidal butterflies, and that day from my life
though I knew—ten more faces remained for the night,
ten deaths for the earth's next day.

After midnight's turned off the light
I can still reach the infinite,
and with my hand, as though I were touching
only a walnut leaf about to fly away,
the cold thigh of a woman who wants to wear a shroud
even on weekdays.

So I slip once more back to my body time
from the midst of timeless hours and timeless years
as blessed as someone bleeding to death in his dream
and smiling, his cheek halfcovered.

(Jascha Kessler)

THE DOOR CREAKED THREE TIMES

You come in. Everything is where it should be:
the keys, the needles, the masks. Gloves.
The hunched beetles.
Ants teeming on the kitchen tiles—
this occupying army.

And the numb roads are the same.
Where's your home? Your country?
Your enlarged wounds?

The door creaked three times
as you came in. The mirror blurred
three times as your arm crashed
onto the table and left its mark
My eyes that cruise the world,
searching for closed lids,
for a dream where there's room
for your body.

Deny this arrival,
this order without happiness.
The crimes of submissive feet.
Women scramble in front of your house,
their eyes speak of a flood.
Of the charm of a catastrophe.

(Nicholas Kolumban)

THE TENANT WHO LIVES IN FIRE

For a student

He wraps himself in flames
as if he were cold;
he sits inside fire
the way he sits at home.
He just invented a country for himself,
this country nobody can take away—
it branches out of his shoulders,
out of the roots of his hair.
It's lodged in his loin-valley,
deep in his flesh,
this resting place
of weary girl temples.
He drives out the heat from within,
his many deaths,
his many raptures.
He climbs the mound

268

of burning haystacks
upward to the vault of the sky,
where there's no exit.

And the bone flute crackles
and sings from under his skin.

You, tenant of fire,
death pilot,
your name evokes dread
in the daylight.
Your name shines vividly
on days when there's no light.

(Nicholas Kolumban)

A BULLET

I could use cool water or strong coffee
that makes one live forever
so I can become giddy.
Maybe the rustling of a shirt, made of snow drifts,
on my skin
or bird claws and iron shavings on my eardrums.
Maybe the lewd hip show each morning
of the woman next door;
the slow rocking of her ocean-going breasts
past my window.
Or the imprint of a brotherly elbow
between my ribs.
A bullet lodged in the window frame.

(Nicholas Kolumban)

THE BACHELOR

My friends murmur in my ears
that I'm not compatible with you, poetry.
They say I should walk out on you.
You're a beauty queen—

both a virgin and a whore.
Your lashes flutter in the wind, northbound,
like wild geese.
I'm the slave driver—
I plod after you even this summer,
drenched with mud.
It's true: there was no street, no house
no fine moss-lined hotel
where you waited for me.
There was no city where you'd fall in my arms,
wall white, staggered by human blood
that rains on roofs.
I did run after you
from Warsaw to Havana
with the memories of tanks, saxophones
and throbbing, wild movies.
I wanted to seduce you
but I only saw your veil of iron,
your veil of ocean.
I think of you with the shame
and rage of a life
that has been half-eroded
like a shoe worn down on one side.
I'm jealous but I don't accuse you.
Go and flaunt your bare, dark loins.
Your feverish thighs that could send
less deserving boys than me to heaven.
Your linden-blossom scent will do for me
and the leaf-veined sky
that your breath drives in my face.

(Nicholas Kolumban)

ÁGNES GERGELY (b. 1933)

Ágnes Gergely studied Hungarian and English literature at the University
of Budapest, and has had a varied career as a secondary school teacher,
besides holding positions with Hungarian Radio, with a literary weekly,
and with a Budapest publishing house. At present she is an associate editor
on the staff of the literary monthly *Nagyvilág* (Wide World). She spent a
term at the University of Iowa, with the International Writing Program, in
1973. She is the author of several novels and works of shorter prose fiction,

and translator of works by James Joyce, Edgar Lee Masters, Dylan Thomas, as well as by contemporary Nigerian poets. Her volumes of poetry include *Válogatott szerelmeim* (My Selected Loves, 1973) and *Kobaltország* (Cobaltiland, 1978), from which latter the following poems are taken.

AUDEN IN OLD AGE

At last the critic Toynbee wrote with ease
of all that flowed from you: the vodka, booze;
replied your friend, the angry Stephen Spender:
to hurt a corpse this great just isn't tender.

You, with your handsome horse's head, chauffeur,
unsightly in the Spanish Civil War,
expert at rhythms killingly collapsing,
at worm-infested loves, at lullaby-lapsing;

walking the razor edge: cliché and terror,
caught in the end in your own verbal tether,
no, not of whisky, not of metaphysic
would it attack: the youthfully ecstatic;

old name, witness to great times you've become,
and no one would now say you'll start it again;
on the banks of the Thames no single oak
walked with you—only anxiety's tavern-smoke.

Rhyming even without, you ancient lion,
losing your good name, lonely over wine,
where even Eliot was no competition,
not the childless, not those with old-age pensions;

guilty or blameless? only you would know;
but while in the corner of an attic window
a candleworth of poem lights that dormer,
there's always one who, seeing it, grows warmer.

(Emery George)

ENCELADO AND THE FACTS

Encelado Sulfato XVII, King of the Cobalts,
awoke one fine morning to the fact that he had no left hand.

"Majesty," said J.P. Brux, M.D., Brooklyn physician
on official visit in Cobaltiland,

"you probably never had a left hand.

"Inasmuch as you did have one, the present *casus* may
give rise to apprehensiveness only if you by chance are
left-handed, Majesty.
    The left hand's indispensability,
on the other hand, is by no means proven; on a trip to the moon,
I should say, you could find that the right ear
may be used as the left hand.
    What is it that the outside world
names a shortcoming, Majesty? A single man
grows a wrinkle at the end of his nose,
walks his dog early in the morning;
people say of him: 'Well, well, why doesn't he have in him
the strength to raise a family.' They don't see that, nights, he
populates his darkroom with well-known elves doing somersaults using
cursing left hands, right legs.
What, just what is the left hand good for?
    Bid welcome, Majesty,
to painless asymmetry,
get used to it, there will *be* no two exits,
for independent is the one who does without,

and independence is readiness, as is the waning of the moon."

"I have no left hand," said Encelado XVII,
King of the Cobalts, while placing the honorarium
in Dr. J.P. Brux's, the well-known physician's,
solitary cigar-pocket. "You, Sir, from Brooklyn,
are an outstanding diagnostician.
    Hey, relatives, courtiers, quick!
summon the Cobalt sorcerer, maybe he'll *say* something."

(Emery George)

272

GNOSIS

Encelado: you and I.
Vapor trail on a cemetery sky.
Geese in file in a forgotten village.
A winter branch.

River under uncut ice.
Walls.
Print in a book that has seen no
paper-cutting knife. In a mouth

a held-back curse.
Two-day-old fingernails.
That "Thank you" never mailed.
People talking: parallel sentences.

Beggar: one who
completes the synthesis of
held-out hands.
Eternal pointer at leaves' ends: "To you."

Entrance examinations' anxiety toward
the future; talent's
crown-stripping labyrinth,
by the time you recognize it—

no time. In vain you thought
of it in good time: when
danger comes you'll leap out of the way,
what's left to the

moment is preempted
by astonishment: "Is that all? . . . ";
there is nothing for it after the fact.
Repeat it and you lose on the gamble.

Don't be history's fool.
Encelado: paranoia, pal.
Fear recognized in good time,
that too is all collapsing: the other wall—

273

beauty, goodness: shards, pence.
In the dark
strength is all that glows.
Get intense.

(Emery George)

UPON A RENAISSANCE CARVING

But Sin on his knees is
more appealing. His monkey-face, shivering ivory
ribs shine with desire. His tempting flat feet
convince me too. The key lifted in heroic
horror, the Greek profile, Virtue's preciously
guarded navel repel me.
    And yet. What retains us
on the agora boiling underneath?
Perhaps resistance with its dubious outcome.
    The frame of a poem, forever empty.
The masterpiece which can't be played over.
    The plant which droops for no ultimate reason.
    Death, duty-free.
Horses' hysterical desire
for freedom, revolutions.
    The dream of St Theresa of Avila in that walled-
up, summer-scalded garden, that once, just
once, at the close of one letter, she will
erect for Him a high and sacred
exclamation mark, tearing all else asunder!

(Timea K. Szell)

SILENCES

    Lips.
Eyes behind your neck.
Closed briefcases.
Vase (what's in the vase)
(what's in the neighbor's
vase)—that is the question.

Conversation on the bus, over
attentive heads:
"And what's new in Nyasaland"—
"In Nyasa it rains throughout the fall."
Meaningful glances.
      Telephone.
The telephone's monolithic silence.
Dogs' eyes alone.
The Laokoon group.
Crying in the evening park
The space for sails (it could
be: a battleship)
ocean, the uncleavable.
Ten thirty. Too late to
call anybody.

(Timea K. Szell)

MAGDALENE

I was a coffin, and I became a flower.
I was a maiden, and they gave me away.
Chastity is a potential graveyard.
Lust is a waning moon.
      I've tried
everything, Savior; your trials nailed
me up; my gate's opening
knew blood, flesh, and bones,
until
      the child-sized moral stumbled
out of me: we must not judge, but
endure. Nails, blood, moon configuration,
cell-wall, covered cheeks, hoes—are but
possibilities. Treason and death screams
are not fulfilments—only further possibilities.
How long will
      the riches of your choices multiply?
How deep are the wells of our patience? Oh, tell me
that time itself is possibility! And that because
I have been abandoned, so am I!

(Timea K. Szell)

275

MÁRTON KALÁSZ (b. 1934)

Kalász was born at Somberek, in Transdanubia, into a family in impoverished circumstances. He attended school at Pécs, and worked in western Hungary in various capacities until 1957. Since then he has been a reader for a large Budapest publishing firm, a member of the staff of *Új Írás* (New Writing), and a visiting lecturer at the Humboldt University in East Berlin. Kalász's poetry is characterized by strict formal discipline and tight language. *Az imádkozó sáska* (The Praying Locust, 1980) is his ninth volume of verse.

MAIDEN

May that lovely young body of yours
leave but its bare trace of you in me!
My admiration's a straggling soldier's,
with just the sense his might be,

but I shant seize you, shant bite or eat
—that flesh of yours all so careless
and grown ripe though still so child-sweet,
where I have never been, even less

hope to come, no, how could I get there
or do anything . . . and all because
I've never felt my own years like this,
stunned by this moment that gives me pause—

it's hardly old age, no, the chaos
of my mere twentysix years
out of which one makes one's own shape,
too soon yielding like this to one's fears

even before one must; and just see me
putting on the cheery expression,
not the lean and hungry face when
conversation turns on passion . . .

Well, what you got from me was this:
a crock of crummy advice, plus
a little yearning selfpity, when
he's the one to redeem me thus,

yes he, even he should live now
for these coming few years in me,
making it worthwhile for him to step out
towards you like a soldier, free.

The nostalgia's as far as I go,
never reaching the words—you see,
the one looking at you's not this young
twentysix-year-old, but only me,

older, tireder, the man virile
enough to note his tension, the wild,
secret explosion in his body
as he contemplates you, child;

but finds he lacks what it takes
to cast his avid net over you,
and hasnt learned the heroism yet
to feel his shame through and through.

(Jascha Kessler)

LEGACY

I don't see my mother dancing—
in my thoughts she still trims vines
sprayed blue with copper sulfate
for her two bags of wheat, eight bushels rye.
I don't know if her young face
was lovely,if the other tenants
admired her dragonfly form,
or if my blond father tethered his horse only
at our cabin in the wild Whitsun ride.
I just see her in the wintry dawn
chopping cornstalks at the stove
or patching sacks in the stilled yard;
I see her at evening in the vineyard
secretly taking flowers for my dead father.
Such memories pour into me,
and whirl me round fiercely now—
my mother, whom none could help,
in the darkness of whose flesh

277

the cancer spread its deadly arms,
who left her son this legacy.
This is not to blame her, not one curse
ever left her lips, I know... Only, poverty
took it all from her vein-roped hands.
Half a day she walked to find me, a hand
at some far-off farm, bringing me potatoes she spared,
spending her scant savings on my studies;
and when I scanned my first lines
at the window something silvery
glowed in her eyes—joy.
And then she was gone, never to see
the first book. I could thrust no money
secretly beneath her bolster, for a dress, for salt—
her bones in the graveyard
moldered to fat silent clay; now flowers force their roots
in summer where her forehead used to be.
And I carry her legacy for good:
on my face the mark of sorrow,
in myself humility's soundless load;
until I die I shall not forget
that world of grinding poverty—
in the field we are walking
like yoked horses together forever.

(Jascha Kessler)

HYMN

Where should I seat you, where
should I place the vase with its shaggy bouquet,
where should I set my favorite books out for you to notice,
what say to you or hand you when you ring,
and will that once-over on the pictures do it,
won't my daughter's little hands
        show up by lamplight,
how to lead round to this splendid carving,
that Roman fragment, these remarkable pebbles,
        my past, my children, my mother,
and that sometimes I madly think
        I've actually invented my own private theory,
though everything you see, chair, rug, mug,

278

stands hidden here in this mean cave, like me,
how suggest my loneliness to you,
    my loves, my misery,
    how reveal it all,
even my naked sobriety, when you're here at last,
    darling future?

(Jascha Kessler)

IMPROMPTU

Like Chopin, not even showing my puss,
letting my loved ones fume through the flat,
letting them hate me in the meadow, where
their outraged cries are braided in the poplars
like ribbons in hair—staying indoors,
writing, discovering short poems,
obstinately poems, neither French
nor Polish: both at last discernible
in them, and then in each other—
where's my country? where my poems
are so wellmade I count myself at home
in their nutshell; in their images the landscape:
poplars shining through, cries for crowns—
so I neednt even trouble stepping out for them.

(Jascha Kessler)

TIME'S NO GRAMMAR

Time's no grammar
rain weeps and falls to make me speak language
to it; May wind
blows through my skull, driving words ahead—
for my heart; the moment limps
spastically
until the sun comes out, and there! your grinding teeth
on the other side flashing some appeal—
watch it! murmur the delicate,
little banners of stucco tumbling in slow motion

279

back onto the walls now: from behind
they sift down on me from everywhere—
autumn's here, silent scraps of paper
mixing with the loud flakes of first snow.

(Jascha Kessler)

UP AGAINST THAT WALL EVERYWHERE

Up against that wall everywhere—
listening to invisible time
in the dark; then, at the core of memory
they set an earth-eating tot
in a dazzling garden; abandon it there—
knowing that I could sit here forever
moved by myself alone, nearly unconscious;
and whatever became of the good instinct, and trust—
that earth-eating tot's tiny palms flitting
so simply from creature to creature; has my consciousness
thus vanquished concept after concept—
the lips shaping, the trance casting words there;
the child seated, unerringly tucking away
the most mixed-up stuff of existence.

(Jascha Kessler)

OTTÓ ORBÁN (b. 1936)

Orbán studied at the University of Budapest, but left without obtaining a
degree, and since 1958 has been supporting himself from his writing. A
voluminous translator of verse, he is also the author of literary and
autobiographical essays, as well as of travel notes commemorating a
journey to India that he made in 1973. Orbán is the author of nine books of
poetry, among them the collected volume *Szegénynek lenni* (To Be Poor,
1974) and, more recently, *A visszacsavart láng* (The Flame Turned Back,
1979) and *Az alvó vulkán* (The Sleeping Volcano, 1981). For his first five
collections he became a recipient of the Attila József Prize in 1973.

PEACE PILLAR

Greek summer!

Softly the sunlight treads the houses as the vineyard
    and in the dense, sweet walls, in the road's foaming white
    floods the inattentive delight of victory.
There sweeps the strong pace of your glance in the white flood,
    in the aerial architecture of faces lifted to the trees' crowns.
So I am, such a windswept brow, such an upraised hand!
So, under the arcades of living days and love,
    in the slow and fine dust storm of days that make it home!
Light carves my profile into stone, as it does law.

What feet and hands!
    What beautiful, awkward motions of creative strength!
What tame bull's necks and cotton-textured white teeth!
How loose a softness of faces turned to stone
    permeates earth and cloud,
    igniting an elusive morning's eternal light!

Full, spacious air! Only this is eternal in us, this hunger!

This strong time, which walks through my bones,
    as does the idea of the first house through matter and rock.

(Emery George)

GYPSY CHRISTMAS

The basalt wolves are howling now,
Herod gallops on a sea of sand;
look how the modesty of grasses rots,
how they point, like the deaf-and-dumb.

Now with an Assyrian stone beard on a gypsy kid's chin,
delirious time perks up its head,
now, mangy with mud, the dogs howl
from nowhere to nowhere, like the gods.

But look: today the livestock exhales marble,
Golgotha-dialect noses the manger;
the Three Kings are bringing a big basket of stars.
Why don't they bring a glass of water instead.

From under snow smells the good-smelling sun,
country of the poor, already bleeding;
on a thatched roof sizzles creation's fat,
spring wades up to its knees in soot.

I'm thirsty, thirsty, for you sentenced me here,
beautiful one, to graze on salt and death;
your poisoned apple is sending me its fragrance,
between grass and sky it's bound to come tumbling.

(Emery George)

THAT STRENGTH

That strength, which behind the tongue
dug out the root of screeching,
hammered a breath-stake into the throat
and chained to it biting teeth;
that strength, which dislocated
from their sockets concentric countries,
where in the cottonlike boughs of blood
weeping sweeps through like a winged ghost;
that strength which, revolving
in mud, as do the dead in their graves,
tossed with its groaning pick-and-shovel rage
a few word-clods across the mouth's rim;
sexton of sounds, in the high seat of
creaking, a coachman in mourning,
hiding in hopeless chatter
perennial hope—that strength
hammers winter's bolt off your mouth,
shakes the stars and dams up the seas,
passes judgment on boughs, pursues birds;
that deathly taste may rot in your mouth:
your torment's embers grow sharper,
the spring of the death of death,
tickles your tongue: it's like flint,
like a knife on a grindstone, it slims and screams.

(Emery George)

A SUMMER ON THE LAKE

The shore seemed suddenly more distant. Or had the other slid close?
It was nothing, though. They just climbed out. Guzzled peaches on air
mattresses.
As in the beginning, the dead sardines glistened
in the divine can's blued sauce. But on Earth's Screen
their broadcast destiny was visible only as a ghost. Location indeterminate,
outlines blurry: the fuzzed picture
of a time not past, not future. Themselves and yet not. Nor was that sour
taste
in their mouths from something more they knew. Whatever could that have
been?
Not even the paltry possibilities that had been theirs:
the same old streets, same old loves. They'd lived where they could live;
and living from day to day's the same anywhere, isn't it? The portables
gushed their silent dance tunes. Illusory lives, glaring darkness.

Afterwards they chatted. And their jokes—oh well, those! Yet their gossip
resurrected the heroic age of their breasts, that downy gold
tingling along their thighs. And Tooth & Claw, Afrodite of personal
mythologies,
stands at the door again: "She could at least have had the decency
not to phone what it was like from my own bed!" The ashes-and-dust
wagon
rattled along, accelerating. They slapped their shanks, haw haw haw.
Their kids dived off the landing: applause, kodak smiles
caught and fixed. Yes, it was they all right: who'd taught
the frigid lady how to enjoy it and the nymphomaniacal model how to
cook—
these commuters on a branchline to heaven. And love's
inane, shiny, pharaonic mask glowed defiantly
on their faces tilted sunward.
And then, shrieking, they were off on their way to Hotel Bones.

Behind them, worn out by leafing through the centuries,
the wind creaked an open window.

(Jascha Kessler)

CERTAIN YEARS

And down came the years one after another
    like black clouds We might as well start like this
It's a straight line from here to poetry socalled
    His The-Beautiful-is-what-pleases-without-exciting-the-passions
                poem
sprinkled with a vision rising out of the kneadingtub
    and naturally a star's always welcome as a raisin
in the universal coffeecake for its light alone shining
    on the sentence Everything seems
okay Image after image 100% of capacity
    And the city where I grew fearfully wise
emerging after the war Lips shut
    eloquent eyes No meat Mother coming home with an empty
shoppingbag And in the pesticidal movie the newsreel's
    black-and-white dazzling with the future's Once-upon-a-time

It's easier this way Not raising the question
    are people to be blamed for what happens to them
And if they are because one way or another they are
    how and for how long? How big's the individual share of blame
the collective? And where do the stories start?
    What conqueror wasn't conquered once?
And which of the conquered wouldn't prefer conquering?
    And who hasn't a good excuse for anything?
But if everything's merely a link in the chain reaching to the primordial cell
    what's this acrid flavor in my mouth? Whose blood?

I recall I'd just stepped off the sidewalk The burst
    nearly flung me against the wall I saw his face
Like mine the scared face of a child He just hangs onto
    the machinegun and mows away Any set of years
anywhere You squeeze yourself into the future between two years
    the barking harrying your heels You know him that
improviser capable of the lesser of two evils
    or an inexplicable good His own prisoner To that extent free
Around the monkeyface blooming on the poster tacked to the Earth
    glory radiating everywhere

(Jascha Kessler)

REQUIEM

He tossed a life preserver to the young castaway in '55
a longish first translation job for the bedhopping ingenu
about this monster poem says he
it's an epic shtik you can live off it meanwhile
he'd remembered me as the precocious monkey

the years then
our years on earth
first shooting then silence is all it is
up close everything seems small-time
but seen from a distance
as though one thumbed through Revelations
between a bloodypimpled sky and burning houses
we both aged
a pair of family men

I have my notion of Resurrection
the angel reeling off the official text
took me for someone else when
he pinned the brass medal
to my rib

the afterlife 1964
raining windy
K's coming from the direction of the Tuileries
his sweater blooming as big as the Czar's Bell
its hem hanging on him like a skirt
he says *let's sit down someplace before we freeze*

Sainted Trinity of paupers
he Julie and I
pooling our francs on payday
laughing as though we lived it up
and we are alive because we've invented immortality
(the kind we can afford)
a bottle of rosé

(Jascha Kessler)

SNOWFALL IN BOSTON

everything began here
and not just with the Tea Party
the rebels scalding the tea leaves (and before that the
taxes) of His Royal Majesty with icy harbor water
and not even on this continent
the city layout testifies to this clearly
with its English brick houses on the soil of Indian huts
and in the John Harvard Library of Cambridge Massachusetts
the name recalling king-beheader Cromwell

the moral is not to be found here of course
it's snowing in Boston
the cars skid on Boylston Street
and under Beacon Hill salty slush is tanning the boots
and the city is slurping the rushing flakes
of heavenly hopes as black mud soup
where the golden dome of the State House flickers

the moral is not an English tailor shop
nor is it the historic buildings
Indian Negro and all sorts of other blood has firmly hardened here
and in Chicago along Lake Shore Drive the highrises line up like launching
pads
guard in the doorway cameras follow the visitor around
and there is always a plane above a certain point of Lake Michigan
and the penthouse bar slowly rotates and the waitress has no skirt on
and New York white collar workers carry with them a five kilo newspaper
in the morning
of the five four is advertisement

so let's not soak in the centuries
let's go warm up in the Aquarium
the *superb infringement* will here not slip out of its *course*
what's more propelling his torpedo-shaped body with streamlined fins
he calmly circles behind the thick glass
and doesn't go hiding his voracius nature in books
as does history red with the blood of ideas

everything began here
and now everything is in vain
as usual it will end elsewhere and as something else
while above Boston there dumbly circles

with bone-cracking teeth in its gaping mouth
blasting aquariums
and untamable the large white shark
Winter

(Timea K. Szell)

JUDIT TÓTH (b. 1936)

Judit Tóth, recently widowed, lives with her son in Saint Maurice, outside Paris. In 1959 she received her certification for teaching in France; in 1957 she won first prize in an international translation competition sponsored by the Hungarian literary journal *Nagyvilág,* and has been a contributing editor of that magazine since. Tóth has published three volumes of her poems, *Tűzfalak* (Firewalls, 1963), *Két város* (Two Cities, 1972), and *A tér visszavonása* (Revocation of Space, 1975) and, more recently, a novel, *Kifutópálya* (Runway, 1980).

FOREST

There was no way out of that forest.
In its pits the barracks-silence
of hours past midnight had come to eternal halt.
Silent watchtowers the trees,
in the piles of leaves the mounds of rubble,
death's stripped-off objects,
earthenware plate, feather tuft, frame for glasses,
jaws—abstraction's
final experiments.
Motionless, the wire knot of branches.
The roots sticking out, like railroad tracks.

In the crater of the washed-out region,
dark houses lounged.
Supine they lay, like swimming plank boards,
concave in the invisible flood.

In their empty basins sand, ashes,
made waves; in the dry basket of bones
wind hardly stirred, just as on chimney walls—

only loneliness, only soot.
Once form, flesh have separated off,
and possibility has blackened,
how desolate the machinery left.
The barren fence of teeth,
nails, stems, wires,
gnashes together on time come to a halt.
It measures something, as do abandoned walls.

In the noise of the invisible earth-cave-in,
animals were running.
Perhaps they were looking for a way out of the forest.

(Emery George)

OUTSKIRTS, AFTERNOON

To penetrate, in the outskirts, through dark doorways,
into tar-smelling alleys, where plaster
breaks from the walls like eggshells.
And the windows like eyes of coal cellars.
And the mirrors of tobacconist-cafés
reflect, in the wild light,
the faces of the day's accused.

And the machine-rattling of afternoon clocks.
The row of highway lamps
like a burning flower garland,
like an electronic outdoor feast in May.

Saint-Maur-des-Fossés, Joinville-le-Pont,
Varenne, Créteil, Maison-Alfort, Charenton,
start dancing in the rattling procession.

(Emery George)

MURMURING

Here the trees are eight stories high.
Around the cable factory
wetness and mud are eternal.
In the unpaved tracks of cars,
puddle water, muddy sediment, paper trash.
But at the base of the house the garden,
like a green nest, is closed.
Pool, willow, fig tree.

In front of the house the thoroughfare.
As if the clatter of eternal freight trains
shook the past, the future.
Creaking steel structures, like towers lying down,
deliver the cables wound on spools.
Time's metal-winged clacking shakes the house.
And flitting steel shuttles are
weaving the bedsheet of growing distance.
      In the billowing linen shroud,
         only the seasons' blood murmuring.

(Emery George)

MASKS

Faces hover on the windows.
They look out, as if out of the depths
of salt mines. They're grayish-white,
just like the shadow of salt on them.
In the flitting of sheets
and pillows they have nothing on.
Crumbling plaster casts,
in their eyes gray, gray infinity.

I look back at them. They look back.

In the vapors of the upper-floor kitchens—
pickled concavities.
In the dugouts waiting, pulverized
to indifference (or fear), forgotten
appetite, habits leaping out front

289

to soothe parts of the day.
Schedule of meals, schedule of
physical needs—garlic, salt, vinegar.
Trees above dried-out meadows sway
the way their arms wave in the open windows.

And the way the flower-patterned
gowns balloon
in the morning draft.
They take to wing in the dark street,
these ball-masks, just like a cry of pain
on the walls, in the open windows.
Winding vines, bell-cup,
drooped poppies, leaves,
their blue-vitriol, sulphur, blood colors
are ashen, as if a Saharan
layer of dust covered them; and yet they rise to dance
as does above them the tired curtain in the wind.

On the heads, hair-curlers.
In a stiff grouping
a charmed flock of snakes
around the face, on the neck.
Medusa? It isn't threatening. It's the hope of
some sort of bliss it advertises.

(Emery George)

FEBRUARY

February. Germination's moment.
The only one, the brightest of the year.
It's still so far—summer is still incomprehensible,
but here, above the asphalt, trembles
its wind-departure, its fast handhold.

Trees in the suburbs,
buzz of traffic—touch
above the roofs, for but a minute,
just like the ice-cold bush of some
fireworks.

Holidays. Market days.
In the Charenton hangar's
butchershop-courtyard-noise,
among the rumblings of
unloadings on a Sunday morning,
the roaring wind brings around summer.

(Emery George)

TWO CITIES

From a tower the stone sea
with its street-ruffled waves.
From here you can see all time past.

They swim past the soft mouths of
wild beasts; grasp them, details of
landscape, unfolding from fog,
turning away. Solve, hug to you
the lost city; both of them,
the invisible
and the visible. What do you do?
What's your job?

The void backs away. The timeless
sun disc of sadness bursts through the fog.
To hammer out the words, to slice out
meaning with bleeding violence,
to separate it all out of thicket, confusion.

Space, like a shark, opens its throat.
Distance's colonnade gleams.

(Emery George)

ROMANIAN

INTRODUCTION TO ROMANIAN POETRY

By Thomas Amherst Perry

A full appreciation of contemporary Romanian poetry is not possible until one has grasped the meaning of the Romanian word *suflet* ("soul, spirit"), a concept of themselves as a people with centuries-old, almost mystical, bonds to one another and to the land on which they have lived since before Dacian times. These are bonds that extend back into an archaic past, and which have produced a mind-set peculiar to Romanians as *rumâni*.

Essentially *suflet* is a perception of life as an "inseparable symbiosis" of nature and man, of a natural setting whose chief components are the Carpathian mountains, its forests and ravines, the soil the Romanian people have tilled for centuries, and the plains on which they have grazed their herds. It is also a multi-plane perception of reality that grasps the simultaneity of the timeless and the temporal, what Mircea Eliade has termed the "camouflage of the fantastic in the daily event," and the juxtaposition of the absurd and the rational—what Ion Barbu has named "the power of grasping a complex range of elements at one glance." But this has been a perspective tempered by Romanian associations with other peoples: Greek, Celtic, Roman, Oriental, Byzantine and, especially during the past century and a half, the modern Western world.

The years between the two world wars saw an intensified awareness of this *suflet* together with a sensitive responsiveness to modern literary developments: Western changes in the concept of reality, symbolism, free verse, and the broadened range of materials considered suitable for poetic use, including so-called "non-poetic" elements. There was also a fruitful exploration of the latent poetic potentialities within the Romanian language and folk materials. Both developments produced a group of writers who set the direction for contemporary Romanian literature, even contributed innovative techniques and modes to the rest of the literary world: Tzara (a Romanian pseudonym meaning "countryside") and Fondane (Fundoianu) introducing dadaism (from the word *da-da*, meaning "yes-yes"), Urmuz anticipating surrealism, Paul Celan (who turned to German), Eugen Ionesco and the theater of the absurd, a contribution to French letters, and Mircea Eliade, with his understanding of the fundamental meaning of myth and his insights into realities within fantasy and the paradox of the timeless and temporal.

This ferment produced four poets who have become major influences on contemporary Romanian poetry: Tudor Arghezi, Lucian Blaga, Ion Barbu, and Gheorghe Bacovia (1881-1957). Unfortunately their work and their influence was halted, first by the war and then by the demands of social realism after 1948. The works of all four were declared "decadent" and all but Arghezi were forbidden publication. Arghezi himself, unable to conform, was silent until 1955, when he wrote some verse with social implications. The published poets too often replaced lyricism with rhetoric, but there were occasional moments of poetic feeling, usually inspired by responses to world events, as in Jebeleanu's *Hiroshima*. The death of Stalin and a more independent course for Romania led to the rehabilitation after 1960 of all these poets, and editions of their verse appeared, making possible their influence upon a new generation of poets.

Arghezi had broadened the range of poetic expression to include peasant language, underworld argot, and imagery from the seamier sides of life. He had also articulated his spiritual quest for meaning in a turbulent world, reflecting his own personal odyssey from Orthodox Christian monk to positions of agnosticism and then atheism, ending with those answered questions about death—a search for spiritual verities with which many of the younger poets could identify. Blaga, proving the roots of the Romanian *suflet* in the archaic past and in the village, employed folk myth and folk symbols in his search for the essential spiritual values in that soul-state. His verse was also characterized by a sensuous imagery rooted in the Romanian temperament.

Barbu, a professional mathematician, utilized the natural polysemy of ordinary Romanian vocabulary for poetic effects, revealing simultaneously different levels of meaning within the single word or phrase. His ideal was a compact poetry of "essences" comparable to the "abstractions [and] associations of mathematics." He was admired for his verbal music, provocative symbols, and for his intellectualizing of the concrete and transfiguration of the commonplace through "poetically descriptive acts of incantation." Bacovia, the most talented of the Romanian symbolists, brought to that mode genuinely native qualities, creating verbal music from words and language rhythms, and expressing a restlessness and a sublime fear in a world of potential disaster—emotions shared by some of the younger poets.

Building upon the groundwork laid by these four writers, the younger poets have created what Professor Don Eulert has described as an "explosion of multiple, exploratory, and self-conscious poetry" in four major directions. Besides revival of the lyrical tendencies found in the verse of Arghezi, Blaga, Barbu, and Bacovia, there is a refinement of Romanian surrealism, continued experimentation under the influence of the latest European techniques, and a "contemplative, refined poetry of silence in the presence of mortality." There is frequent use of fragmented imagery, parable, or myth—sometimes historic and sometimes invented—and traditional folk symbols to communicate spiritual truths. In much of the verse of the young poets the old *dor* (tone of sadness, longing, found especially in the folk *doine* and similar to the quality found in the blues of the American black) reasserts itself.

TUDOR ARGHEZI (1880-1967)

Tudor Arghezi (pen name of Ion N. Theodorescu) is one of the most important figures in modern Romanian poetry. As a young man Arghezi went through countless odd jobs, was ordained a deacon of the Romanian Orthodox Church, worked as a journalist, and was imprisoned during both world wars. Among his best-known works are *Cuvinte potrivite* (Fitting Words, 1927), with its shift from symbolism to primitive speech patterns, and *Flori de mucigai* (Blossoms of Mold, 1931), under the influence of Villon and Baudelaire employing seamy images and underworld argot. His more recent titles, *Frunze* (Leaves, 1962) and *Călătorie în vis* (Dream Journey, 1973), show the poet on his sustained quest for meaning, for a sense of Romanian archaic past, and for poetic means of portrayal of feeling for the weak and underprivileged.

TESTAMENT

I'll bequeath no goods to you when I am dead,
Only a name upon a book instead.
In the rebel evening which comes through
From my forefathers down to you,
Past precipices and deep ravines,
Up which my folk have clambered on hands and knees,
And which await your climbing while you're young,
This book of mine, my son, will be a rung.

Enshrine it at the head of your bed,
It spells out the oldest charter you have had,
From serfs in sheepskin coats burdened only
By bones which flowed down years and into me.

So that for the first time we might exchange
The hoe for pen, the furrow for inkstand,
Our old folk laboring among the oxen

Stored much sweat over countless eons.
Out of their tongue with which they called their herds
I've woven into my verse fitting words,[1]
And cradles for the masters yet to be.
I kneaded them for a thousand weeks,
I turned them into dreams and icons.
From tatters I've made buds and shining crowns.
Distilling honey from pent-up poison,
I strove to keep in full its potency.
I've taken insult and spun it fine, with care
Sometimes to charm, sometimes to curse or swear.

From the hearth, ash of people dead and gone
I have shaped into a God of stone,
A lofty bourn with two worlds at its rim,
Keeping high watch on your duty to him.

All our bitter grief and mute pain
I heaped upon one single violin,
Which the master heard, and to every note
He thrashed and kicked like a new-stuck goat.
Out of slime, festering pustules, and mildew
I have brought forth beauties, values that are new.
The whip, so long endured, returns as words,
And slowly punishing, it thus redeems
The living spawn of everyone's misdeeds.
It sets the hidden forest bough aright
As it reaches out into the light,
Bearing on its end, like a clump of warts,
The fruits of grief from past eternities.

Reclining idly in her cozy nook,
The princess suffers in my book.
Words of fire and words forged by the mind
Are married in my book, and intertwined,
Like red-hot iron embraced by tongs.
What the serf has written, the master ponders,
Though unaware that buried deep within
Simmers the stored-up wrath of my ancient kin.

(Andrei Bantaş and Thomas Amherst Perry)

[1] This line supplied the title of Arghezi's first major volume of verse.

Note: This is Arghezi's poetic manifesto.

PSALM

I ponder you in clamor and in silence
Tracking you through the course of time, like game,
To see: are you my much sought-after falcon?
Should I kill you? Or kneel down and pray.

For faith's sake or the sake of denial,
Stubborn I search for you, and uselessly.
Of all my dreams you are the loveliest
And I daren't shake the sky to let you fall.

As if reflected in a flow of water
Sometimes you seem to be, and sometimes not;
I've glimpsed you in the stars, among the fish
Like the wild bull when he is taking water.

And now, us two alone, in your great story
I stay to match myself again with you,
Without my wanting to emerge the victor.
I want to touch you and to shout: 'He is!'

(Peter Jay and Virgil Nemoianu)

GULF

white temple over the gulf
of silence and of ages
beyond horizons beyond
the soul's being and its searches

a mirror never stirred
by oar or boat or
wind it is a place
that's not in heaven nor on earth

deserted bay window
reflects the cloister gloom
the blaze diminished
to vacant ashen hew

(Stavros Deligiorgis)

297

ION BARBU (1895-1961)

Ion Barbu (pen name of Dan Barbilian, a distinguished mathematician) was of Wallachian birth. Although he died (in Bucharest) at the beginning of the sixties, his work, along with that of his exact contemporary Lucian Blaga, has been of profound influence on the younger generation of poets. His mathematical specialty—from 1942 on he taught algebra at the University of Bucharest—is said to have determined the form and style of his poetry, the main thrust of which is hermetic. His only book of verse is *Joc secund* (Second Game, or, Counter Play, 1930); its title establishes a view of reality that juxtaposes a subsurface swirl of movement with modulations of the "game" of existence above water. Other poetry, which Barbu published in journals, is determined by esotericism and by folk symbols. An enlarged edition of *Joc secund* appeared posthumously in 1966. Barbu has also done an incomparable version of the opening two acts of Shakespeare's *Richard III*.

COUNTER PLAY

In time inferred, the deep beneath this calm crest,
Thrust through a mirror into a freed empyrean,
Cutting through above the drowning country herd
In swirls of water a second, shadow game, a purer dance.

Oh latent nadir! The poet raises up a sum
Of scattered harps lost in the inverted flight
And a song pours forth: hiding as the sea
Hides the jellyfish when they stroll under verdant bells.

(Thomas Amherst Perry and George Preda)

DIOPTRIC

From high on the spectrum, the prisms ponder
The saturating of a sign, foliate.
Like choice wine, the caulicoles redden,
Though the sun sets at the rim in mourning crape.

Nearer. The eyes squint fixedly
At the leaf vibrating like a drum,

298

At the crown of the letter, bramble-like,
Heavy in the low flutter from the hearth.

The chamber, curving in the gossamer dream!
—Tidied up by old women, it passes by,
Litter shaped into cones, immured within Tomes;
From embers, confirmation of a day.

(Thomas Amherst Perry)

TIMBRE

The bagpipe sear to meadows, the flute to highways dry,
Voice the sundry sorrows with worse or better ease...
But the rock in prayer, the earth stripped by the breeze,
And the vast wave betrothed to heaven will wonder—why!

A song, a song capacious it should be, to vie
With the surfy silken rustling of the salt sated seas;
Or the angel's paean above the apple-trees,
As from the male rib rises Eve's wispy trunk on high.

(Dan Duţescu)

THE DOGMATIC EGG

The dogma: And the Holy Sprit
moved over the waters

This melancholy people is also given
The sterile egg, as food,
But the live egg, a baby in its tip
Is created to hold up to the sun.

Like the old world in crystal
The innocent new egg—
He swims in the thin lime—
A wedding palace and a tomb.

Of three Indian satins is the bed
In which the snowed white is sleeping.
So languidly, so closed in,
Like a beloved body dissolving in your dream.

But the embryo?
From very high up
From the plus pole
From where the mud
Of earth is beyond reach,

He gently grants,
Masculinely,
The white in the hyaline,
His full kiss.

*

Forgetful man, irreversible,
You see the Holy Spirit turned sensitive?
Like then, today is just the same:
Tiny worlds observe the dogma.

So you can see the Holy Spirit at the arches.
Watchful at the clean live waters,
I bring you now this symbol-egg,
You plain, forgetful man.

Not the red-dyed egg,
Insatiable stupid man,
It's an egg with a baby that
I want for you, as a present, at Easter:

Raise it towards the sun and know!

And first of all thrill yourself
With that yellow Turkish coin—
A clock without its minute hand,
The only scribe of the time of death
For egg and world. Thrill yourself
With the yellow, necessary clock...

The whole forehead of death is there.

In the yolk,
To gnaw the rich white,
Duration draws a circle in us.
Just like that—the dogma.

Once more: the egg of the sterile one is identical,
But don't sip it. You break up a wedding.
And don't put it under a brood hen, either;
Leave it in its first peace,

Because the whole doing is guilty;

And holy—only the wedding, the beginning.

(Donald Eulert and Ştefan Avădanei)

LUCIAN BLAGA (1895-1961)

Lucian Blaga, a distinguished philosopher as well as a poet, was born in a country village in Transylvania, the son of a priest. His volumes of verse: *Poemele luminii* (Poems of Light, 1919), *Paşii profetului* (The Prophet's Footsteps, 1921), *În marea trecere* (In the Great Passage, 1924), *Lauda somnului* (In Praise of Sleep, 1929), *La cumpăna apelor* (At the Watershed, 1933), *La curţile dorului* (In the Courtyard of Yearning, 1938), *Nebănuitele trepte* (Unsuspected Steps, 1943), *Vîrsta de fier* (The Iron Age, 1940-44), and *Mirabila sămînţă* (Wondrous Seeds, 1960), introduced his lifelong concern with spiritual experience, usually within the context of village life, with the archaic past, and with a sensuous enjoyment of nature. Blaga displays a fondness for sleep metaphors and for "splinters from myth."

IN THE MOUNTAINS

Near the hermitage midnight discovers
creatures asleep on their feet. The spirit of wet moss
walks through the ravines.
From the east come butterflies like owls
to search fires for their ashes.
By firtrees' roots, near the curse of hemlocks
a shepherd pours earth
over lambs killed by the forest's powers.

Passing over the ridges,
the sheepfold girls—bare shoulders rubbing the moon,
their adventure supernaturally imbued
with dust brightly stirred from the disc like a swarm.
Yellow horses gather their salt of life from the grass.
Smouldering under trees God is becoming smaller
so that the red mushrooms
can grow under his back.
In the sheep's blood the forest's night is a long and heavy dream.

On four deep winds
sleep penetrates old elms.
Under the shield or rocks, somewhere
a dragon with eyes turned toward the Pole Star
dreams of blue milk stolen from the pens.

(Peter Jay and Virgil Nemoianu)

TWENTIETH CENTURY

Machines whir underground. In the unseen above towers,
intercontinental electric sounds.
On housetops antennae feel the spaces
with new tidings, in unknown tongues.

Blue signals crisscross streets.
In theaters lights scream, the freedoms of the individual are extolled.
Downfalls are foretold, words expire in blood.
Somewhere they are casting lots for the shirt of the vanquished.

Archangels, come to punish the city,
have lost their way in night clubs, their wings now scorched.
The white dancer runs through their blood as, laughing, she pauses
on the tiptoe of a leg like an upturned bottle.

High up, at an altitude of one thousand meters, to the east
stars tell their tales through fir-tree boughs,
while in the dead of night the wild boar's snout
unearths new springs.

(Mihail Bogdan)

THE SOUL OF THE VILLAGE

Little girl, put your hands on my knees.
I think eternity was born in the village.
Every thought is more silent here,
and the heart beats more slowly,
as if it were not in your breast
but somewhere deep in the earth.
Here the thirst for salvation is healed,
and if your feet are bleeding
you step on a bank of clay.
Look, it's evening.
The soul of the village is fluttering by
like a shy smell of cut grass,
like a fall of smoke from thatch eaves,
like the tumble of kids over tall tombs.

(Don Eulert, Ştefan Avădanei, and Mihail Bogdan)

HERACLITUS BY THE LAKE

By green waters the footpaths gather.
There are stillnesses hereabouts, heavy and abandoned by man.
Be quiet, dog, testing the wind with your nostrils, be quiet!
Don't chase off the memories that come
crying to bury their faces in their dust.

Leaning on logs I guess at my destiny
from the palm of an autumn leaf.
Time, when you want to take a short cut,
which way do you go?

My steps resound in the shadow,
as if they were some rotten fruit
dropping from an unseen tree.
Oh, how the voice of the well has grown hoarse with age!

Any raised hand
is merely one more doubt.
The sorrows yearn
for the earthy mystery of clay.

I throw thistles into the lake from the shore,
with them unfolding myself into circles.

(Peter Jay and Virgil Nemoianu)

ANNO DOMINI

Night has entered the town, unheralded,
and snow falls again under grey hours.
On the cathedral's eaves, medieval
ghosts of the woodland languish.

The clock's tolling stirs a bat
from the long sleep in which it had settled.
The ash of angels burnt in heaven
falls in snowflakes on our shoulders, on the houses.

(Peter Jay and Virgil Nemoianu)

VIRGIL TEODORESCU (1909–1987)

Virgil Teodorescu studied modern literature at the University of Bucharest,
worked as a journal editor and as president of the Romanian Writers'
Union. Some of his early poems were published, under a pseudonym, in the
avant-garde review *Bilete de papagal* which was edited by Tudor Arghezi.
Some of his titles are: *Blănurile oceanelor* (Ocean Furs, 1945), *Semicerc*
(Semicircle, 1964), and *Repausul vocalei* (The Vowel's Rest, 1970). His
translations cover Paul Eluard, Nazim Hikmet, Mikhail Lermontov, and
Romain Rolland, among others.

GOODBYE

I could call to my support a dozen trees
            in the prime of their age
and several household objects which have proved
            their usefulness in excess
but I don't want to make an issue far from me the vain
            wish to pull it through with the help

304

of the trees though I wouldn't doubt
for a minute the feeling of sympathy
they surround me with
I do prefer however to let them grow all the time
I am going down
bit by little bit it's not very noticeable
better this way than otherwise
I am going down without complications without ceremony
I am going down like an aged mountain
blasted at by sands and by winds

(Stavros Deligiorgis)

AMELIA ERHARDT

if you could know that roads are oranges
    marsupials dropping Athena's laws of iron
if you could know how to open the cat's eye and flee right through it
    your hair whipped on by the squat sea wind
    towards the ashen cafes stifling the age of trees
if you could know how hot it can get when you are covered with keys
        how pained the sound of the aged metal in the illustrious
        medieval keys locking the eye of the beloved woman
        on nights of love and rain until its rusting
        neck veins would burst like the flight of a faraway horse
        homesick poplars growing in imperial halls under mahogany easy
chairs
        a new dynasty coloring its salt cellars mauve
if you had known how to flee if you had known how to sleep
    on the side of roads of no end

    like some aerial plants or oranges

(Stavros Deligiorgis)

AT NIGHT THE JELLYFISH UNFOLD THEIR VEILS

I'd like us to open the cabinets and let the resin of firs out
And let your cheek flow forever
So that I may speak of the people on Mars
Of earth's tongues entered deep in the ocean
Or of amulets made from reeds
It never occurs to anyone to open the cabinets
So that all liberties be allowed to us
And your cheek be a locust again
The gloves again be stuck to the flesh
Your absence be the untouched glass
The thin veins of milk be a safe road
Along which three wild horses might go abreast
And all the sacred places of your body
Be more poisonous arrows.

(Donald Eulert and Ştefan Avădanei)

ARM IN ARM

I call you, we'll see the snails fall and break
Arm in arm
We'll look at the long decay of telescopes
In the explorer's eyes
at the threads of snow among teeth
We'll look at the sledges falling in ravines
at the way seal fat burns too slowly
Arm in arm
We'll look at the shanks of mares
as at some flowers I'd like to hand you
together with the hook for fishing corpses
at the way your flesh slowly, almost invisibly, diminishes
I call you to the core of the just-split coal
that no longer glimmers like cobwebs
nor like the dark circling of the octopus dizzying its prey
we'll look at the conspiracy of household objects
against the dead shells on the point of becoming objects
at the way their windows open when they no longer have
                    room inside.

(Donald Eulert and Ştefan Avădanei)

EX VOTO

Come. Or if not remain where
the green-colored triumph can be seen
the defeated fruit of clear berries
dead of famine
gathering into themselves heaps of sun-rays
there where whitish like smoke
in air purified by frost
rises a feneant condottiere
a hollow full of bats towards which
my ship floats in a snowing
of splinters with adjacent flights
and fluctuating continent corridors
fixed like a lupus on the retina...
Come in the lighted polygon
crossed by those broken birds
that forgot how to sing when they hatched
and fish linked by a membrane
tied into the diaphanous flesh
disfigured by the pressure of wool
conflict of the wonder-beings
that are born in air, lay eggs in air
and in a marble whole are two.

(Donald Eulert and Ştefan Avădanei)

EUGEN JEBELEANU (1911–1991)

A winner of the prestigious Taormina Prize—granted previously to such
poets as Dylan Thomas, Tristan Tzara, and Giuseppe Ungaretti—
Jebeleanu is undoubtedly one of the most exciting, as well as most
productive, poets of his generation. His more than twenty volumes, since
1929, include the epic work *Surîsul Hiroshimei* (The Smile of Hiroshima,
1959), *Cîntece împotriva morţii* (Songs against Death, 1963), and *Hanibal*
(1972); his work and vision range from the epic to miniature lyric
meditations on history, as in "Hannibal," presented below. Jebeleanu is a
translator of poetry from six languages, including Hungarian and Turkish,
and is a much-translated poet in his own right. He is a corresponding
member of the Romanian Academy of Sciences.

HANNIBAL
To Giancarlo Vigorelli

Nobody had what he had
his superb pride
and the elephants
their pads crushing the vertebrae
of these Alps white with fear.

He stepped, barely heard, over rocks
and was heard on the moon, and
nobody had ever seen the waving
stone trumpets of such beasts.

And they didn't defeat anybody.

(Donald Eulert and Ştefan Avădanei)

ŞTEFAN AUG. DOINAŞ (b. 1922)

Ştefan Aug. Doinaş (pen name of Ştefan Popa) was born the son of a peasant family in western Romania. After studies in literature and philosophy at Cluj (where he heard the lectures of Lucian Blaga) Doinaş taught in his native village for a number of years. Later he worked on the editorial staffs of various periodicals, including that of *Secolul 20*, a Bucharest monthly devoted to world literature. He began to publish poetry regularly only after 1964, and in 1968 won the Writers' Union Award for his collection *Ipostaze* (Hypostases). His important books since then have been *Alter ego* (1970), *Ce mi s-a întîmplat cu două cuvinte* (What Happened to Me in Two Words, 1972), and *Cai în ploaie* (Horses in the Rain, 1974). Comprehensive collections of his verse appeared in 1972 and 1979. He has also published three collections of essays on poetry, and translations into Romanian of such poets as Dante, Hölderlin, Mallarmé, Valéry, and Gottfried Benn.

THE MAN WITH EXPLODED EYES
to V. Nemoianu

In mud lodges, below, I have conversed with the Mothers,
    down in a garlic-plot where the blaze with bulging

308

bulbs strangles the chasm. There
I stayed, without mercy—and here I am, blind.

In his withered mouth the god masticates his offspring.
Who plucked me from his jaws? Clattering they
closed behind me, those mighty fangs, and
doomed me—saved—to an unnatural death.

With eyes open I was going down in wonder through the gray
strata, bearing digested words on my tongue,
as others still carry in a soul's wrinkle
a whiff of sown grain to the Styx.

But I was not heading with the boatsman for the valley of
[shadows
Alive rather, but less than the Silent Ones, a tribe
lurking for me in the depths, I was passing submerged
landscapes, I was ploughing bones and frontiers.

By mud and blaze thresholds they waited for me. Only
my step's teardrop, were it to grow like flax
behind me, could tell with blue flowers
where the sunny peaks left me and vanished from.

In blaze lodges, below, living like a breath of raw
bagpipes just in the belly as wind in barrels,
with their vision turned on itself the Terrible Ones
stared at me, receiving me in their country.

Swarming as hearts: to no one, to everyone, outside the body,
themselves having no such thing; like uddery sows
gasping, scanned by their tender sister,
the ravening Vicinity; instead passionately

loved by Places of unsteady aspect; forever preparing
them their form and pattern. Only rabid
packs have such a heated panting, but
not the flame, too, trickling from the snout.

With their tongues they loved me; tore off my shoulders;
[my heart
stopped, frozen with fright, nausea-fraught
at being a looker-on to non-being, and
to my birth, and that of races in thousands.

309

Also to the turgescent udder from which the morning stars
    emerge, as from a sling, slimy-feathered birds,
        and the seas—salted glands, and the mountains
    like boils slowly petrifying their pus.

The visibles, ah! they are cooling. But in their entrance hall
    the original fire is a beast: it bites the creature's
        potency to be the world's witness. They bit
    my eye as though it were a flowering lily.

This I declare, and a smell of singed sows is spreading.
    All that is poured into moulds here, I have seen
        there fermenting, and I have time only
    to smile blandly at the Frowning Ones.

So, and not otherwise, do I feel lightning-struck. The darkness
    only protects those ravenous for light of day.
        But he who is saved from the demesne of flares
    shall wander the world with exploded eyes.

(Peter Jay and Virgil Nemoianu)

THE ASCENT

Certainly the mountain in front
is manageable: some like to take
it head-on...
            But we prefer
(to combine business with pleasure)
a more roundabout route, which also
feeds the lofty taste for the beautiful
landscapes and coolness and resting-places,
and the murmur of elms trilling to the sky.
It's wonderful to set out together
like good comrades, and slowly, slowly
to spread out on the way...The heavier ones,
naturally, will drop behind. But it's good
to feel them breathing down your neck
ever more poisonously, as they gallop
round the bends. There are some who won't
leave, their steps keeping pace with ours.
What you need is a slippery rock,

a subtle movement to make them
fall down the precipice. Sometimes
glances, gestures, words are most useful:
you turn to one of them aiming
contempt strained through an eyelash, a small
superior smile—and at once
you see him hurtling downward as if
hit by an unseen thunderbolt.
Frequently, though, it happens that
the temptation comes from themselves; the moment
they slip on a tuft of grass, they utter
a thin cry, make an indescribable
face, which begs you to press your heavy
foot on their chest, with the sole
of your boot stepping on the cramped leaf
of the hand wanting to hold on.

Odd that their glance, in which
the tall mountain is like a pin's
head, instead of flashing like a sword,
only throws back astonishment.
Hawks catch the screams in their beaks.
And around, there's such a boisterous bustle
of predators! But rain will come,
wiping all traces away...
                              And on we go,
stooping a little, but with satisfaction
in noting that we, who make it,
we all belong, in bearing and feature,
to the same stock: the glorious race
of winners who feel at home
on the crowns of their fellow men.

The view all round is magnificent.

(Peter Jay and Virgil Nemoianu)

THE WORDS

Yes—somebody has lived here.
One senses it in the brute smell
and in the wind prowling around
these sweet shells. A misalliance
with space was the rough breath
that passed by here and stopped
amazed that the world's entire
texture should tremble to its marrow.
Hear? The ghost of a Sarmatian
sea still pushes its waves toward us,
and angels faithful to those thresholds
rush at our mouths when we talk
in sleep. Of course, there must have been
some kind of lunatic tribes here, hordes
of riders fed on raw scents, or
crusaders bent on concocting
their God of sound . . . Everything
bears witness: through the old sore
of a consonant, through the cleft
of vowels, that all the world's lovers
—before falling into each others' arms—
came here to strip off their shadows.

We got here late: this area
is now a reservation of silence.

(Peter Jay and Virgil Nemoianu)

MOMENT OF PARTING

The distance between us becomes lightning.
And birds fly through the hot cinders,
with livid shrieks. And the alien
passing rapidly, like a runner
cannot, alas, feel that his chest has broken
a slender tape that was shining,
with desperation, between our eyes.

(Peter Jay and Petru Popescu)

THE WOMAN IN THE MIRROR

From behind the mirror, a row of mirrors
throws you to me through centuries—look:
you take off your Greek gown, part
with the mottled wool garb of the peasant;
like fumes the crinoline falls down
over your hips and the veil disappears;
from a tight dress you produce a shoulder
like the ivory of bitter fall's quinces;
and you turn, your hot breasts burdened
with kisses, on your lips the same words
that led the saints into temptation
and, having stolen halos, they remain holy for ever;
with a hieratic gesture, walking tiptoe,
you come close and sweep your breath over my face
hotter than the roar of the docile lion,
lighter than the wind, purer than the flower;
and deep, from the somber well of time,
where only the flame and the stone speak,
a language ignored by goddesses on Olympus,
you call me back, as Cleopatra called the svelte serpents.
Today's bite wanders over the eternal
wound like a sword over unbroken mail;
and I lose you, and I call after you, while you move
with the same smile into a new mirror.

(Donald Eulert and Ştefan Avădanei)

PYTHIA

glib page
           the nearsighted priestess
is trying to decipher
my writs
dreams of holding
the ball of air but the letters
faster than her stained finger
like the black waters sharing
in the moon's dilapidation
pass it
hand to hand behind her back

313

and again the name and the point
reversed of this terrorized game
on verso laugh
between
crazed lines

(Stavros Deligiorgis)

VERONICA PORUMBACU (1922-1976)

Porumbacu was a poet and an author of prose works; she left, besides her
poetry, three volumes of prose and a large body of translations from the
English, French, and German. Born in Bucharest, she died there in the 1976
earthquake. Among her principal volumes of poetry are *Histriana* (1968)
and *Mineralia* (1970).

THE PUPIL

Answer me, friends:
   what good would it be
   to wish for an ephemeral house
   to adorn?
   Like gracing
   a train compartment,
   on wheels
   gnashing into nowhere.
   Ahead,
   a row of blind semaphors,
   fires
   long since extinguished:
   space without time,
   a neutral time,
   perhaps a well
   into which I am falling
   and I shout, and I shout, and I shout.
   Above,
   the eye of the day, still white,
   no bigger than a moon,
   than a gold coin,
   than the pupil:
   ... minuscule snake eye.

(Marguerite Dorian and Elliott Urdang)

ANESTHESIA

And suddenly
  the silhouettes hush
  irremediably
  like the drowned in winter
  under the river's ice,
  like saints, vertical, enigmatic,
  behind glass.

(Marguerite Dorian and Elliott Urdang)

FARTHER AWAY

What would they say to me when,
asleep,
I add myself
to the anonymous crowd,
my bones,
farther away from me than
the old rocks of the moon...

(Marguerite Dorian and Elliott Urdang)

FROM WHAT ONCE WAS

From what once was, from the crazy nights
with horses' clatter through the dust of the Milky Way,
with the rope-walker's step on the line
                                between night and day,
with the cry of the moonbeam cutting like
                                diamond the window pane,
with the storm in a single maple leaf,
and the northern lights in a single flower of ice—
is left only a thin haze,
dream within dream, or maybe
for my blood with which I wrote
transparent runes on the sky,
only insomnia with which I ask
tribute from memory.

(Marguerite Dorian and Elliott Urdang)

NINA CASSIAN (b. 1924)

Nina Cassian is a prose writer, an author of children's books, and a musician, in addition to being an eminent poet, prominently of physiologically aware and erotic verse. For many years she was an editor with some of Romania's premier literary publications, and a leading figure in the Romanian Writers' Union. She is the recipient of a number of prestigious literary prizes. Among her fifteen volumes of verse the later, and now better-known, are *Ambitus* (1969), *Chronophagia* (1970), *Requiem* (1971), and *100 poeme* (1975). She has also been a highly-regarded translator from the German and the Spanish.

PUBERTY

Alone on the empty beach.
I stood naked on the lonely shore,
The sea naked and empty.

And a dragon-kite flew over
with black head red eyes murmuring
black, red paper words.

Alone he saw how naked I was,
red and black in the palm of summer—
and he fled into the sun to forget.

(Brian Swann and Michael Impey)

TEMPTATION

Call yourself alive? Look, I promise you
that for the first time you'll feel your pores opening
like fish mouths, and you'll actually be able to hear
your blood surging through all those lanes,
and you'll feel light gliding across the cornea
like the train of a dress. For the first time
you'll be aware of gravity
like a thorn in your heel,
and your shoulder blades will ache for want of wings.

316

Call yourself alive? I promise you
you'll be deafened by dust falling on the furniture,
you'll feel your eyebrows turning to two gashes,
and every memory you have—will begin
at Genesis.

(Brenda Walker and Andrea Deletant)

THEN

First the crow is heard
Then the wild pigeon.
Then someone washing, wringing out
cheap rugs and water dripping
in the sink. Then
a screech of wheels.
The oblique wind. Then
the broom whispering. The sea
pulling away. One hears
the door of the afternoon sky
swing open smoothly.
The leaves begin a tiresome
trip. One hears
the bird of sleep. Maybe too
the walk of a cat, a kitten
in her mouth as she crosses the backdrop.
Again the pigeon. Then
water in the sink. Then
the sea far away. Then
autumn.

(Laura Schiff and Virgil Nemoianu)

THE YOUNG VAMPIRE

At first, shy, he coiled himself around
my neck, with melodious volutes,
so that my whole neck was enveloped
in the bracelets of that melody,

317

and I was almost taken with his ugly
head triangular with eyes askew
and the sound of his frail bones.

Then, at the first bite,
I felt great relief.
My blood throbbed, aware of jumping,
thinned, into an alien maw.
Its color became purer
and I was emptying, as if of sin.

Later, I was greatly thinned.
The flutterer sat tense upon my neck
and was drinking, drinking me.
Ever more freely his wings flickered,
his eyes burned like two letters
—but I could not read onward.

(Marguerite Dorian and Elliott Urdang)

THE BLOOD

Ah, I still remember well that pain!
My soul taken by surprise
Jumped about like a hen with its head off.
Everything was splashed with blood, the street, the café table
especially your thoughtless hands.
Strewn about, my hair wandered
like a monster among glasses,
coiled around them as if around suspended breaths
then danced, vertical, whistling,
and fell, executed, at your feet.
Ah, I remember well that I smiled savagely,
disfiguring myself so as to look more like myself,
and that I cried only once,
long after everyone had left
and the light was off and the blood on the tables
had been wiped clean.

(Marguerite Dorian and Elliott Urdang)

318

ION CARAION (1923–1986)

Poet, anthologist, and translator, Caraion is considered the spokesman of the mature generation of Romanian poets, and the most important existentialist voice heard at present. He has a very large body of work and has frequently been translated abroad. Among his more recent poetry collections are: *Eseu* (Essay, 1966), *Necunoscutul ferestrelor* (The Unknown One of the Windows, 1969), *Cîrtiţa şi aproapele* (The Mole and Its Kin, 1970), *Cimitirul din stele* (Graveyard in the Stars, 1971), *Munţii de os* (The Mountains of Bone, 1972), and *Poeme* (1974). Caraion has also translated voluminously from American and French writers; in 1968 he was awarded the Writers' Union Prize for his translation of Edgar Lee Masters's *Spoon River Anthology*.

MEMORY

It was war and leprosy, men dying in the town—
contorted, hideous
black vases of flowers
verdigrised at dusk, in the waters' depths.

Gentlemen, gentlemen,
I'll never get done writing—
my circus suffers from directed promiscuity
—it was war and leprosy—
every horse had to be born a second time,
since someone had slipped into the levy the explanation: "encore".

Our pacts were being closed with artifices and compromises,
we knew to respect every jar in the exhibition
and for every junkshop
we reckoned out several portions of national dignity and a few new...
utopias.

Everybody stop! We're living moments of great solemnity.
One can't forget such moments... preserved in formalin or lime,
here and now, no agrarian system
should ignore them.
The government sees to everything and is preoccupied, as you all know!

I was fond of that century with energy and machine,
and I said:

if not, what we hold is just a caroming ball.
Every horse had to be born a second time,
the buttons whelped, the motorized moon gave off the smell of gasoline.

I'll never get done writing...

Men stretched their arteries in the air
like beautiful, useless signal-bells
and then waited for 40 days
but the prophet had run out of miracles and died.

I felt for him, odd like a reindeer:
rotting quietly among cranes
he had been the last rich man in that museum
turned several times in place.

I am inviting you, gentlemen, to this break-up of the flu
and offer you my vats of water boiled into the immobility of the masses
whom I no longer can imagine sitting up
facing the machine gun—that desperate woman strangling her children
in the fields and screaming.

I invite you to the banquet where you can't eat a thing
because all the corpses have begun to rot,
where translators no longer have work; and our remaining reaction
is to smooth down this belly partially dispossessed of its navel.

I am at any rate still fond of that century
with its dirty stuffed and mounted breasts,
where the prophet, removed from the window display,
was, like a gas lamp, quietly rotting.

How curious is the carcass in the dusk!
The children stretched out on bags in the yard
were eating garbage and wiping their mouths—
it was war and leprosy, and men were dying in the town.

(Marguerite Dorian and Elliott Urdang)

DEORNAMENTATION

No one discovers anything for you, alone
you discover the miracles you can believe in.
All other miracles have died a thousand times
in a thousand people
and no one wonders about them any longer
but to lie again, and again.

This weariness was needed, needed
these solitudes to surround me,
to be without splendor, to hear
as before the symmetrical air—
... not to be able to tell anything.

To hear
the gunfire, the spoken sadness
of the city
twice ornamented in red.

With barrels
stained green by absence
needed the wound, needed the wind
unlimited and chaste. And the tear
forbidding sleep.
Nothing but the shadow. Under butterflies
another sky deceives.

No one discovers anything for you, alone
you linger
in the verdigris of grass among skulls and crayfish.
A thousand birds walk without a magus—
and then it is silence, it is weariness,
always there is someone who won't be gone enough.

(Marguerite Dorian and Elliott Urdang)

ALWAYS, IN MARCH

Even uprooted, still their flowers
going on blooming, the trees
looked around: centuries old and adamant.

Always in March,
racing the torrents, driving essentiality to despair,
the trees were overwhelming.

They mauled each other, they mingled
like will-o'-the-wisps,
like planets
or like a bag of enormous silences
mingling with the time inside the bag.
With time disfigured by immobility.

Better: like some electors of light
invaded by their own restlessness and doubts.

And then the seeds would burst forth.

(Marguerite Dorian and Elliott Urdang)

THE ENVELOPING ECHO

A woman crossed the park and laughed.
With hoop, with kite, with sling,
the children ran about the sky.
A woman crossed the park and sang.

That fall was like a bunch of grapes.

(Marguerite Dorian and Elliott Urdang)

BY THE LIGHT OF THE BRANCHES OUTSIDE

Bursting with their own wisdom, the pods were opening.
The leaf was falling, befuddled, tiring the air.
A bird thrashed above the mulberry tree.
Like a Saint Gheorghe at the circus, in autumn,
returned from the future of my body,
I was drawing the unseen.
The days were pounding into silence like a pillow.
Blacksmith and tender of the fire,
I look at the mountain, cathedral

with white candelabra of larch,
under which sat Eurydice, crying.

After midnight, when the clowns philosophize,
I look into the mountain.

(Marguerite Dorian and Elliott Urdang)

BITS OF SILK

I was friends with all the solitudes.
I lit up lamps in wanderers.
Evenings I drank some tea, or not even that.
Paths into the past have narrowed—
and here's oblivion.

All is as once was:
things which I cannot give a name.
Girl with fairy tales in your hair,
let's give up remembering.

In fall the circuses would leave.
The old woman sold marjoram to us.
Darkness favorable to pawnshops,
still the wind makes somersaults and butterflies.

Once you were showing me a squirrel big as a potato
and we unwove at the figment's whim.
People know something they are not telling me.
How's the water into which you shook your hoary glooms?

Through grass and wet seasons,
the ashes mingle their saints.
The evening came like a dog from the mountain,
to lick your warm hands.

Still you are my love and still
I hear the moon snake through the walls.
Oh! If we had stayed in fancy only
like the wars on panoplies...

Life was always as life should not be.

(Marguerite Dorian and Elliott Urdang)

WINTER

The huts under the drifts have frozen;
the wells in the frost—outside.
It snows. Polar night.
Through his sleep a bull hazily dreams
his last spring's grass.

(Marguerite Dorian and Elliott Urdang)

ARCHAIC HIPPOCAMP

Fate is like a face
behind
the wash hung up to dry
a girl with pale eyes
among cows and shirts
on the sky in the grass
reads
from
Tolstoi
the morning's liquid wonder broke
the crickets out of war and peace
jump through the soul rotten underneath
the field with cauliflowers and walnut trees enters me
things open and thirst enters their mouth
space enters vanished happenings

(Marguerite Dorian and Elliott Urdang)

NICHITA STĂNESCU (1933–1983)

Nichita Stănescu was born in the oil city of Ploieşti. He received his degree
in philology from the University of Bucharest in 1957. After his literary
debut in 1957 simultaneously in *Tribuna* and *Gazeta literară,* he rose
quickly to prominence in the sixties. Among Stănescu's most important
volumes of poetry are *11 Elegii* (1965), *Oul şi sfera* (The Egg and the
Sphere, 1967), *Necuvintele* (The Unwords, 1969), *În dulcele stil clasic* (In
the Sweet Classic Style, 1970), *Cartea de recitire* (The Book for Rereading,
1972), and *Starea poeziei* (The State of Poetry, 1975). Stănescu's work has

been translated into a number of European languages, both Slavic and Western, and he has been a recipient of many literary awards. Besides his own poems and essays Stănescu has also translated from the Yugoslav poets Vasko Popa and Adam Puslojić.

SECOND ELEGY, GETIC

to Vasile Pârvan

There was a god in each tree-stump

If a stone cracked, a god
was hurriedly brought and put there.

As soon as a bridge collapsed
there was a god, lodged in the empty place

or on roads, once a hole appeared
in the asphalt, there was a god placed in it.

O don't let your hand or foot get cut,
deliberately or by accident!
For they will at once put a god in the wound
as they have done everywhere, all over,
they will place a god there
for us to worship, because he protects
everything that splits from itself.

Be careful, warrior, don't lose
an eye, or they will bring
a god to put in the socket
and he will stick there, petrified, and we
will stir our souls in praising him...
And even you will rouse your soul into motion
praising him, as you would praise strangers.

(Petru Popescu and Peter Jay)

HOMO PLANGENS

Through the state of this soul
go possible existences—
a mortal net tearing up all,
leaving but absences.

I leave the place this guise,
put it frozen to sleep. It will stay
in the pure suspension of those
who have not come, who are on their way.

Nothing remains itself
for more than a singular moment.
I plead not guilty, living
in a world never constant.

Turning only a thought away
from you, I detect nothing there
behind me later; testifying to this
on my knees, raising words for no ear.

The stone in my heart decays.
It remains the center of shifting
edges. Surely I am a word
which has gone to sleep on your tongue.

(Petru Popescu and Peter Jay)

MARINA

Nothing is left. Nothing can be recognized,
presumed or supposed.
Behind me, a horse made of nothing, still grazing
on the nameless and invisible grass.
Ahead, the view becomes clearer,
the glance suspended with fishing lines,
occasionally jerked by a fish made out of light,
which you eat, then throw the skeleton away:
or plant in the place of a fir tree on a mountain
submerged in the sea; where you fall through
long and purple clouds made of drowned bodies
until those bodies cover you.

(Mariana Carpinisan and Mark Irwin)

WINTER RITUAL

Always a cupola
another one always
taking on a halo like a saint,
or only a rainbow.
Your straight body, my straight body
as during a wedding.
A wise priest made of air
is facing us with two wedding bands.
You lift your left hand, I lift my left arm,
our smiles mirror each other.
Your lovers and my lovers are crying
tears in syllables like Christmas carols.
They take pictures as we kiss.
Lightning. Darkness. Lightning. Darkness.
I lower one knee and fall on my arms.
I kiss your ankle with sadness.
I take your shoulder, you take my waist,
and majestically we enter the winter.
Your lovers and my lovers step aside.
A ton of snow overturns on us.
We die freezing. And once again, only the locks
adorn our skeletons in spring.

(Mariana Carpinisan and Mark Irwin)

ADOLESCENTS ON THE SEA

The sea is covered with adolescents
who, standing, learn how to walk on waves,
resting their arms on the currents,
making use of the sun's strong rays.
I lie, watching from the shore's
perfect angle, which they seemingly approach.
An infinite fleet. And I wait
to see a wrong step, at least
a slip up to the knee, as they advance
in the gauze of the breaking waves.
But they are slender and calm, now
having learned how to walk on the waves, standing.

(Mariana Carpinisan and Mark Irwin)

ENGAGEMENT

He didn't dare come home
Loneliness adorned his wife,
who became his bride
She held, on her left ear
a two-winged butterfly
and was trying, but not able to cry;—
its spirit had a scent like lavender,
and the air was like all birds flying!
She was lying and white,
on her finger, the falling wedding ring
was like a comet, shining.

(Mariana Carpinisan and Mark Irwin)

EVOCATION

It was a melancholy, a kind of sadness
a kind of emptiness traversed by an imprecise desire,
a kind of decaying of words into images
representing sierras, maybe even Sierra Leone
because it sounds more remote, more inaccessible.
It was a kind of darkening by which you are refused,
a kind of minor off-handedness, as might be
that of dance,
from which you are rejected only by a gesture
of indifference.
It was a melancholy, a kind of sadness
of a child lost on a childless day,
a kind of game that had happened at a different age
which you can hardly remember,
a kind of game you didn't play
though this happened long ago.
It was like a mail plane of an older type
passing over the eternal mountains
in a space of solitude, passing solitary
and leaving its shadow in the shape of a cross,
that keeps shrinking over my suddenly bare
chest, my shirt torn.

(Irina Livezeanu)

328

SONG

Only the present instant remembers.
What indeed was, is not known.
The dead keep giving each to each
their names, their numbers, one, two, three...
There is only what will be,
only the happenings that haven't happened,
hanging from the branch of a tree
unborn and half ghostly...
There is only my body, now wooden,
the last, an old man's, of stone.
My sadness hears the unborn dogs
barking at passers-by unborn.
Oh, only they will truly be!
We, the inhabitants of this second
are a lithe nocturnal dream
running anywhere, a-thousand-legged.

(Irina Livezeanu)

ION GHEORGHE (b. 1935)

Ion Gheorghe, a graduate of the University of Bucharest, came into poetic
prominence in the sixties. In a solid series of poetry collections since the
mid-sixties, Gheorghe has displayed interest in a poetic integrity that seeks
its roots in European and Latin American modernism. A strong line of
development is seen in his most important works: *Nopţi cu lună în Oceanul
Atlantic* (Moonlit Nights on the Atlantic Ocean), which won the Writers'
Union Award in 1966; further, *Zoosophia* (1967), *Vine iarba* (The Grass
Comes, 1968), *Cavalerul Trac* (The Thracian Knight, 1969), *Mai mult ca
plînsul* (More Than Tears, 1970), *Avatara* (1972), written after a visit to
Cuba, and *Megalitice* (Megaliths, 1972), in which the poet identifies
himself as an archaeologist, reminiscent of the achievement of Charles
Olson.

MOONLIT NIGHTS ON THE ATLANTIC OCEAN

Moonlit nights on the Atlantic Ocean!
I looked through the peep-holes of the universe and died.

329

Oh, God, how I slept, like a white river pebble;
the water passed and I stayed.
As I sleep now, like the mast of a ship;
the water stays and I pass . . .

Moonlit nights on the Atlantic Ocean!
The storm and its waters threw me to all corners of the horizon
only seeds are turned and winnowed, like this
before being buried behind plows.

I am mixed with water and sun, with planets and dangers
as if somebody took me for a seed for the next planet
in fact I see the edges of that endlessly falling world,
places where the moon and the sun climb one after the other,
like golden paths in a clearing empty of grass:
this is the only sign that somebody walks there—
the endless path of a sad light,
and only ships, like stones thrown at random,
like large fruit fallen from a nonexistent tree,
change their path and darken the space around, leaning on
                    nothing.
Abyss, ocean, destruction of dates and holidays,
nothing reminds me of land any more.
I put my head into the aqueducts of blood and look;
it's dark to the top of the most unimportant branches,
the wolf cub sleeps, his muzzle by the thigh of the trainer
he dreams of his parents' pack and runs after sleds at night—
from below, from his parents, the smell of blood burns his
                    nostrils
and suddenly he tears the curtain that separates him from the
                    flesh of the young mistress,
up to his neck he plunges in the blood-pond and leaps
                    through his prey
striking his head against the heads of the parents he had
                    forgotten.
Moonlit nights on the Atlantic Ocean!

To dream that wheat grew from under your nails
to hear the desert-fire wheels of saddened planets,
to dream that you felt like going out to plow and you died
                    with ague;
to rise really ill with earth;
to dash into the telegraph booth and see two handfuls of earth
torn out by nails from a park in Copenhagen,

to look till you fall asleep at the few leaves of grass on it
to smell with a flower loamy breath of the clods...

Moonlit nights on the Atlantic Ocean;
all my rivers with their beds changed,
sleep comes like a bird ill with wandering—
at the hour when it used to leave me...
Oh, God, how I slept, like white river pebbles;
the water passed and I stayed!

As you slept on my arm, on the trunk of a young apple-tree
cunningly placed your head on a block;
night came and I beheaded you till day;
in the morning you cleared like springs after running over
                                    millwheels;
now your shadow stretches quietly at its known place,
sleeps on my arms, while you walk around alive
when you begin to sleep on the shore with your head under
                                    your wing
I imagine you rising from the stones of sleep and from grass—
you get up seven hours earlier
as if every day you were born earlier . . .

Moonlit nights on the Atlantic Ocean;
I looked through the peep-holes of the universe and died.

(Donald Eulert and Ştefan Avădanei)

CATHEDRAL STONES

On sacks of cement they eat bread and watermelons,
on freshly laid stairs, blowing sand off newspapers—
on the shirts they took off lie the stones,
silent by ones and twos;
from each seed shaken onto their knees
two glass leaves sprout;
then wind blows against the walls until they grow.

As many stones as peasants in the city;
if young they eat a loaf a day,
half a kilo of tomatoes each;
they cross the desert of words' birth,

331

they sleep on the wooden doors of the circus,
in the stink of water coming from the gladiator's nostrils;
the consul shakes out his shirt in their presence,
he washes from his neck the hair from a fresh trim,
and asks this assembly of the charred palm-tree;
Quousque tandem abutere, Catilina?...

The stone cares little about someone else's patience;
it longs for the village until blood revenges,
escaped from the bed of a narrower channel;
they sleep on stacks of cucumbers and in shacks,
still living on two loaves from home.

Big stones the peasant and his sons,
he stole the horse's stomach,
and boasts in the world about the burdens he's chewed:
he whistles, and where he whistles the house grows;
he puts a leaf of basil over the key-holes...
own them in good health, says his head
and he goes back to the village once a week,
dragging by its leg the sack of bread.

A young man comes out of the cornfield with his coat over
                    his head,
he scares the girls and women hoeing.
The fool marries at the beginning;
his wife can hardly move for the stone in her belly;
builds houses where he goes;
he comes back broken-backed with sacks of food;
he's brought rice and sugar, toys and bay leaves
and again pours peasant stones into his woman...
from now on who can run away?

River stones the peasant's children; water flows,
some lucky slab reaches the cathedral tower,
sits with the clock on its knees, and in the middle of the door
his father carves the sad dreamer Uta,
puts babies in her arms and forgets her in the bell-tower;
he who does not sleep prays, but the majority chew
unconscious that above everybody's head
the woman cries, crazy with the absence of her man
driven into the world to fetch bread.

(Donald Eulert and Ştefan Avădanei)

THE UNICORN

Come, Sire, come,
First you were and then you weren't
On the dome of Saint Sophia
Watch the heron sleeping—
How long should a kingdom last?

The Gate in the Balkans
Catches fire every hundred years ...

Come, Sire, come,
First you were and then you weren't
With four horses they tore your body apart:
Set one piece out toward the east
To guard the border
Set the other in the west
As a stone to mark the boundary ...

The remaining two
They threw away
Under the icons of the church
Out of longing and waiting
They became grape vines
With clusters ripened into wine.
Whoever drinks it
Will bear a heavy sword ...

Come, Sire, come,
Give me a hand with five brothers
To free you from the four horses ...

Ion Voda understands and waits
For his hand to be called.
When stars turn and turn over
Fire and ashes of August,
He mounts the horse with a horn
And goes out on his quest for luck.

The horse called Unicorn
Knocks with white legs on the gates of the world.

(R. Loring Taylor)

MARIN SORESCU (b. 1936)

Sorescu was born in Bulzeşti, a village in the province of Oltenia. He received his degree in philology from the University of Iaşi in 1960 and afterwards moved to Bucharest, where he made his literary debut in 1957. Sorescu is equally well known as poet and playwright, has written essays, children's books, and a novel, and has translated from the work of Boris Pasternak. He himself has been translated into English, French, Italian, and Spanish, and has received a number of awards, including the Romanian Academy Prize for his play *Iona* (Jonah), and the Gold Medal at the poetry ospide in Naples, both in 1970. Among Sorescu's major works of poetry are: *Moartea ceasului* (Death of the Clock, 1967), *Tinereţea lui Don Quijote* (Don Quijote's Tender Years, 1968), *Unghi* (Angle, 1970), *Suflete, bun la toate* (Odd-Job Soul, 1972), *La lilieci* (By the Lilacs, 1973), *Ocolul infinitului mic pornind de la nimic* (Around the Small Universe Starting from Nothing, 1973), *Norii* (Clouds, 1975), *Descîntoteca* (Disenchantmenteque, 1976), and *Sărbători itinerante* (Itinerant holidays, 1978). Sorescu's art is frequently a personal, yet personable, anti-poetry. *Don Quijote's Tender Years* appeared, in translations by Stavros Deligiorgis, with the Corycian Press (Iowa City, 1979).

ADAM

Adam paced the footpaths of paradise in a
depressed mood not knowing what he lacked still

god then goodnaturedly fashioned Eve from one
of Adam's ribs which pleased the firstborn so well
he felt for the rib right next to it
he noticed his fingers stunned by firm tits and a
sweet rolling butt that looked for all the world
like a well turned note Eve had just shown up
in his path compact in hand and painting her lips
that's life sighed Adam
and brought forth another one just like her

thus as soon as the official Eve turned around
or would go shopping for gold myrrh and frankincense
Adam would bring another houri forth into light
from his intercostal harem

god took notice of Adam's unbridled spell
of creativity called for him gave him a
godlike rimming and kicked him out of paradise
on grounds of surrealism

(Stavros Deligiorgis)

A SEGMENT

I shall begin to forget you
Northwards
You will forget me
Southwards.

First I shall forget
Your soul.
Which will take me
Less than three days

Then, I don't know,
The taste of your hair,
The matter superiorly disguised
In the form of eyelashes.

You will forget me maybe
The other way round.
The hardest for you will be
To give up your past
Full of patterns
Like flowered chintz.

At any rate,
We will forget each other pretty well.
Then, we will wash our hands.

(Dan Duţescu)

CARNIVAL

Let us exchange thoughts
Tree, for I don't even know your name.
And with your thoughts
Give me all your leaves
To lay on my hands,
And on my eyes, and on my brow.

In the end
There will be a beautiful
Farewell carnival
And everybody will wear
His festive masks.
I myself want to appear disguised simply
As a green tree.

(Dan Duţescu)

CHESS

I move a white day,
He moves a black day,
I go forward with a dream,
He takes it in the war.
He attacks my lungs,
I think for a year at the hospital,
I make a brilliant combination
And win from him a black day
He moves a misfortune
And threatens me with cancer
(Which proceeds, for the time being, like a cross),
But I place a book in front of him
And make him draw back.
I win more pieces from him,
But, look half of my life
Is taken, and sits on the sideboard.
"I'll check you and you'll lose your optimism,"
He says to me.
"It doesn't matter," I joke,
"I'll castle my feelings."

Behind me my wife, my children,
The Sun, the Moon and the other fans
Tremble for each of my movements.

I light a cigarette
And continue the game.

(Donald Eulert and Ştefan Avădanei)

SHAKESPEARE

Shakespeare created the world in seven days.

On the first day he made the sky, the mountains and the
                    ravines of the soul.
On the second day he made the rivers, the seas, the oceans
And the other sentiments—
And gave them to Hamlet, to Julius Ceasar, to Antony,
                    Cleopatra and Ophelia,
To Othello and to others
To own them, they and their successors,
For ever and ever.
On the third day he gathered all the people
And taught them tastes:
The taste of happiness, of love of despair,
The taste of jealousy, of glory and so on,
Until all the tastes were used up.

Then some types came late.

The creator patted their heads compassionately,
And told them the only thing left for them was to become
Literary critics
And deny his work.
He reserved the fourth and fifth days for laughter,
He let the fools out
To caper and somersault,
And he let the kings, the emperors
And other unfortunates have fun.
On the sixth day he solved some administrative problems:
He concocted a tempest,
And taught King Lear

The way he should wear a straw crown.
Some waste was left from making the world
And he created Richard III.
On the seventh day he looked around for anything else that
          needed doing.
The theater directors had already filled the earth with
          billboards
And Shakespeare thought that after so much hard work
He deserved the chance to see a performance himself.
But, first, since he was tired beyond words,
He went to die a little.

(Donald Eulert and Ştefan Avădanei)

FRESCO

In hell the sinners
Are capitalized to the maximum.

From women's heads are pulled out
With tweezers,
Barrettes, bobbypins, rings, bracelets,
Fabrics, and bed linen.
Afterwards they are thrown
In the bubble of cauldrons,
To watch that the tar
Doesn't boil over.

Then some
Are turned into lunch pails
In which reheated sins
Are carried to the residences of retired devils.

The men are used as well
For the hardest labors,
Except for the very hairy ones,
Who are spun anew
And made into rugs.

(Irina Livezeanu)

338

CARBON PAPER

At night somebody puts on my door
A huge piece of carbon paper,
And everything I think appears, instantly,
On the other side of the door too.

They're rushing toward my dwelling
The curious of the whole world,
I hear them come up the stairs,
Taking the stairs on their soles
And placing them back again
On their way down.

There are all manner of birds,
Watchdogs for the moon,
Roads in passage,
And old locust-trees,
Suffering from insomnia.

They put their eyeglasses on their noses
And read me excitedly,
Or threaten me with their fists,
Because about everything I've come up with
Exact theories.

Only about my soul
I know nothing,
The soul which is always eluding me
Between the days,
Like a bar of soap
In the bath.

(Irina Livezeanu)

HISTORIOTHERAPY

Whenever I'm sleepless,
Before going to bed, at night,
I take an historical atlas
With a drop of water.

339

And waiting for it to take its effect,
I follow with my finger the frontiers
Of the Hittites' empire,
But barely a moment later
I have to resume the operation,
Because in fact the Hittites' empire
Is the empire of the Egyptians
Oh no!—Of the Assyrians...
Of the Medo-Chaldeans...
Of the Persians...

If such is the state of affairs, I think
I myself could go to sleep
Quietly.

(Andrei Bantaş)

IOAN ALEXANDRU (b. 1941)

Alexandru was born in the vicinity of Cluj-Napoca. He studied philology at
Cluj and Bucharest, and classical languages (Greek and Hebrew) in West
Germany under the auspices of a Humboldt Fellowship. He took his
doctorate from Bucharest in 1973, and is now on that university's faculty
for Oriental languages. Besides his poetry he has done translations from
Pindar, a scholarly edition of *The Song of Songs,* and a grammar of
Hebrew (1975). His books of verse include *Viaţa deocamdata* (Life for the
Time Being, 1965), *Infernul discutabil* (Debatable Hell, 1966), *Vămile
pustiei* (Tolls of the Desert, 1969), *Imnele bucuriei* (Hymns of Joy, 1973),
and *Imnele Transilvaniei* (Hymns of Transylvania, 1976). A selection
entitled *Versuri* (Poems) appeared in 1977. Alexandru's poetry reflects his
scholarly and mystical interests; humility in the face of nature, the dialectic
of joy and a sense of tragedy inform his mature work.

MAN

With the man behind me
I carry a door—its wood
brutally shaped, wet, and heavy and scratched.
It is as if all the absurd battles of the world
with horses, arrows, and helmets

had taken place on its broad slab.
The damp plain harasses us from behind.
We are at dawn and we slither down a slope.
Last night it rained and the ground
is slippery under our heels.
Our white shirts are filled by the wind;
we seem to carry a corpse among billows
of unmoving wheat. The man should
go in front. He is taller and
as he plods the slope heavily most of the load
burdens me and there is swelling
in the veins of my neck and I want very much
to swear a mighty oath. Who are you, man?
When I slip, the man behind
laughs with his mouth like a broken wheel.
When he slips, I would crumple
if the door, discarded scrap wood,
did not find a sort of balance.
We are going slowly, God knows where.
When we cross the streets,
we silently take off our hats
and listen—the fate of the world erupts.
Then I ask him no more; I know where
to go, and among the narrow trees
we shift the door slantwise to make it
through. Big drops of
rain fall from the leaves as we shove
aside the branches.
They are cold under the collar of my shirt,
on my shoulders and eyebrows, and I am hungry.
Straight from my sleep I was taken
on this incomprehensible job
and you, mother, will never know
where your son has disappeared without leaving a trace.

(Andrei Bantaş and Thomas Amherst Perry)

PRAYER

Let the rain come
over the emptiness of the nest in which the egg fell
from the hand of the Universe
under the seething wings,
over the first colt's trace in the field's eardrum,
now when the fingers of doubt
mercilessly swell above us;
through their cracks we hardly sip in
long flashes of lightning that gather the clouds
like herds of swine
under cracking whips.
Through the sieves of the gluey air
that holds its belly tight on the bones of light
so that rain may come:
with knife unsheathed from the earth
let it divide day from night in equal parts,
let it draw gates of appalling rust
between my house and the poor world to come:
let the rain come
to gauge the space in the depth
between the apple-tree's root and the owl's root,
to chafe wounds on the eyes of my window-pane
where daily a cemetery is born;
let the rain come
to pierce open with the awl the udder of the queen-bee
in the desert of honeycombs,
to disperse with arms the herd of lions
on the green hills of hunting.
Let the rain come—aerial homeland where eagles
      hibernate
and the limpid stone is death-conscious.
Let the rain come.
The women in men turn northwards,
and the blood in me again and again
smells of earth.

(Dan Duţescu)

MISERERE NOBIS

This terrible heavy caterpillar
This dumb mouse. This locust
With horse's head and burning feet
That crunches the planet. This caterpillar
Without hearing. This creeping creature
That darkens the universe with its shadows.
This petrified ring at the world's foundation
This poisonous caterpillar which chars
The air and dries the dew of seeds.
This enormous snake terrorizing
The crossroads. This blossomed rot
In motion. This inadequate mould.
This owl overripe with age
This stifled slug. This
Hermetic worm. This spider
Dragging its cross on the hill's horizon
Under the lash of lightnings and in the barking of dogs;
Jeered by all races without sleep and parents
Without shadow and cover. This wretched stone
Reckless, this humbled servant of noman
This mild caterpillar of darkness one fine day
Yes, this terrible caterpillar will become
A butterfly. A saintly and pure butterfly
A white ethereal butterfly, an invisible
Butterfly, a vast butterfly, otherwise
It's impossible. The original butterfly,
That is the butterfly mother,
The butterfly source in which is gathered
The whole race of butterflies that is to come.
The butterfly eye, the butterfly-dream
The butterfly madness, the butterfly son
The butterfly spirit, The Butterfly Father for ever.
Agnus Dei—the butterfly hymn, butterfly
Without robes, the butterfly triumphant
Risen from the tomb.

(Peter Jay and Virgil Nemoianu)

DESCENDING

On those black plowed fields
with the sky dark to the west on one side
and stone-black to the east—
In that northern pit
where you arrive dead-tired and crying,
stands our house
fenced with blood.
In the well's draw-bar,
my crucified father;
the thin lever is my young mother
jointed to his gnarled wooden arms.
And from their ancestors flow
the eternal springs. Time
makes us one with the earth.
In the light of the lantern on the beam,
at night the three babies in white shirts
burn through windows of stretched skin.
And in a corner,
a strange box,
full of days and nights
that a cosmic hatchet drives into our heads
like rusty spikes.

(Donald Eulert and Ştefan Avădanei)

ANA BLANDIANA (b. 1942)

Ana Blandiana was born in the Transylvanian town of Timişoara. She
went to high school in Oradea and received her degree in Romance
Philology and Literature from The University of Cluj in 1967. Blandiana
made her literary debut in 1959 in *Tribuna* (The Tribune). She has been a
frequent contributor to *Amfiteatru* (Amphitheatre), *Contemporanul* (The
Contemporary), and *România literară* (Literary Romania). She has served
as editor of *Viaţa studenţească* (Student Life) and *Amfiteatru,* and as
librarian at the Institute of Fine Arts. Blandiana has traveled abroad in
Europe and the United States. In 1973-74 she took part in the International
Writing Program at the University of Iowa. Some of her work has
appeared in English and Hungarian translation. Blandiana's major works
include *Persoana întîia plural* (First Person Plural, 1964), *Călcîiul
vulnerabil* (Achilles' Heel, 1966), *A treia taină* (The Third Sacrament,

1969), *Cincizeci de peme* (Fifty Poems, 1970), *Octombrie, Noiembrie, Decembrie* (October, November, December, 1972), *Poezii* (Poems, 1974), *Eu scriu, tu scrii, el, ea scrie* (I Write, You Write, He, She Writes, 1976), *Somnul din somn* (The Sleep in Sleep, 1977), *Cele patru anotimpuri* (The Four Seasons, 1977). Blandiana has won literary prizes from the Union of Young Communists (1968), the Romanian Writers' Union (1970), and the Romanian Academy (1974).

FROM A VILLAGE

All of us come from a village
Some directly, others through parents removed
Some directly, bearers of folktales and lore,
Happy ones able to return any time.

Others, through parents removed, wanderers
On branches of blood estranged and thin,
Instead of pastures and seeds and horses they know
The alcohol of the word distilled from these.

I want to retrace the steps of my parents
I want the village with the sound of my tears,
The path through the cornfields I'd know in my sleep
And the reentering of speech into things.

To enter the cemetery with fallen crosses
And recognizing my steps akin to theirs,
Let the ancients get tangled in the roots
Murmuring words of passion in their sleep.

But no one tells me where I'm from. Only evenings
Walking down boulevards, I feel myself close to
The metaphysical doormen near entrances of flats
As near gates, on settles, in a village.

(Irina Livezeanu)

SONG

Leave me, autumn, my green trees
Look, I'll give you my eyes.
At dusk in the yellow wind
From the kneeling trees I heard cries.

Leave me, autumn, gentle skies.
Flash lightning on my face.
Last night in the grass the horizon
Was trying to slash itself.

Leave me, autumn, aflight the birds,
Chase away instead my steps.
This morning the sky wrung
From the skylarks wails.

Leave me, autumn, grass, leave me
Fruit, and leave behind
Bears not asleep, storks not gone,
The hour full of light.

Leave me, autumn, the day,
Don't weep smoke into the sun,
Turn me into night instead,
As I'm turning now.

(Irina Livezeanu)

DANCE IN THE RAIN

Let the rain embrace me from my temples to my ankles,
Lover, behold this dance so new, new, new
Night hides the wind in murks like a passion,
To my dance the wind is an echo.

Up the ropes of rain I climb, tie myself, grab
To make the link between you and the stars.
I know you love my frenzied, weighty hair,
You're taken with the flames from my temples.

Look until your gaze is touched by the wind, at
My arms like live playful lightning rods—
My eyes have never sought the earth,
My ankles have never worn shackles!

Let the rain embrace me, let the wind unravel me,
Love my free dance fluttering over you—
My knees have never kissed the earth,
My hair has never tossed in the mud!

(Irina Livezeanu)

LINKS

I myself am everything.
Find me a leaf that bears no resemblance,
Help me find an animal
Which doesn't groan with my voice.
Wherever I tread the earth cracks
And the dead that bears my likeness
I see embracing and procreating other dead.
Why so many links with the world,
So many parents and unnatural descendants
And all this crazy resembling?
The universe haunts me with thousands of my faces
And my only defense is to strike myself.

(Michael Impey)

347

YUGOSLAV

INTRODUCTION TO YUGOSLAV POETRY

By Vasa D. Mihailovich

The range of poetry in this Yugoslav section covers four areas—Serbian, Croatian, Macedonian, Slovenian—and, for reasons having to do with changes in literature since 1948, for the most part developments in poetry during the past thirty years. This does not preclude representation of a distinguished elder like Edvard Kocbek, nor, of course, a realization that many of the contemporary Yugoslav poets find their precursors and, in some instances, teachers, as far back as the turn of the century. At that time, in all South Slavic literatures but especially in the Croatian and Slovenian, there appeared a new literary movement, referred to as the Moderna, which not only brought forth a number of major poetic voices, but also revitalized poetry in a modernistic direction. As in other countries of Eastern Europe the heavy influence of Western poets, especially of the French symbolists, was unmistakable. The Moderna left an indelible stamp on all South Slavic poetry written during the initial two decades of this century.

Many critics would agree that the only truly worthy rival of the Moderna in recent years, in point of its total impact, was the struggle after 1948 between the "realists" and the "modernists," which to all practical purposes the "modernists" won. The two sides argued on the by now familiar premises, and these ad hoc terms are meant to refer to adherents to positions rather than to theory. However, it would be a mistake to assume that in the struggle the fronts were always clearly drawn, especially along political lines. In fact the controversy was confined to artistic matters; the entire argument was carried on on both sides by sincere supporters of the system. By 1955 the struggle had reached its climax; victory for the modernists was assured by the arrival of a second generation of writers. Two new poets who played a decisive role in this victory were Vasko Popa and Miodrag Pavlović. The principal tenet of the modernists—the vital importance of form-conscious experiment and sophisticated approaches to the materials of art—is, of course, no longer seriously questioned in any of the socialist countries.

Vasko Popa is considered to be the finest living Serbian poet; his poems show an intensely original imagination. There is a kind of primitive, genesis atmosphere about much of Popa's work; in it, each inanimate object has to be looked at and examined again. The poems of the "Little

Box" cycle show this incisively. By contrast Miodrag Pavlović (b. 1928), whom we could not include, is concerned with form, and prefers to seek out the historical continuum which goes back to Byzantium and to the ancient Slavs. The center of gravity in Pavlović's work lies in excavating the emotional and philosophical legacy of Serbian culture. In some ways this poetry finds a kinsman in the work of Ivan. V. Lalić, although Lalić seems far more concerned with contemporary man. Nevertheless, as "The King and the Singer" shows, a penchant for archaic imagery and for a hint at allegory is present even here. One of the youngest poets, Matija Bećković (b. 1939), again an ''outsider'' vis-à-vis our efforts, is possibly unique on other counts as well. His socially conscious poetry, characterized by antipoetic images and by humor, places him side by side with East European moderns like Holub, Herbert, and Różewicz. Bećković may represent a watershed in Serbian poetry.

In the first postwar generation of Croatian poetry, Vesna Parun and Jure Kaštelan occupy prominent positions. Primarily a poet of love, Parun combines sensuousness and compassion with the rich texture of her spiritual intuitions. At the same time there is also a dark streak in her poetry and, as her numerous collections show, ultimately she is a realist. Kaštelan has brought a new, specifically elegiac tone into Croatian poetry, especially in his war poems. In them he laments the dead and reflects on the fate of those who survive. To the new generation of poets belong Slavko Mihalić and Milivoj Slaviček (b. 1928). Mihalić, a neo-romantic with meditative inclinations, attempts to overcome the absurdity of life with an ardent belief in the humanistic role of the poet. Slaviček is somewhat more rationalistic, and places stress on nonconformism. His poems, possessed of the ease of prose, are not devoid of emotion, and have their own kind of intensity. Slaviček at his best has the ability to bring out the absurd details of everyday life and to give them poetic luminosity.

The three poets most instrumental in the foundation and early development of modern Macedonian poetry are Aco Šopov, Slavko Janevski, and Blaže Koneski. Šopov published his first book, *Pesni* (Poems), in 1944. His effort to move the fledgling poetry toward more personal expression touched off a heated controversy, but it also started Macedonian poetry toward a new level of sophistication. Šopov's work is pronouncedly subjective and intense. Janevski is somewhat less subjective, and also bolder in choice of themes, not to mention more original in their testament. Koneski had done pioneering work, not only in poetry but also in scholarly, especially in linguistic efforts. His poems show a somewhat limited range—the Macedonian landscape and people's reactions to it—but in intensity in poetic approach and treatment he far surpasses his colleagues. This is seen especially well in "The Wheat," not only a strikingly sustained metaphor but a suggestively socially oriented poem as well. The two exact contemporaries Mateja Matevski and Gane Todorovski are two further poets who, albeit with one poem each, worthily represent this young literature.

Among Slovenian poets Edvard Kocbek has written sparingly, but his poems reveal a completely contemporary spirit, which places him close to someone like Šalamun. Gregor Strniša, a unique, highly articulate voice, has spoken in an easily recognizable, accessible and yet original, style. His reaction to fear and alienation is expressed through metaphors and dream sequences, cyclically arranged. There is a desire here to mythologize, not far removed in spirit from that of Vasko Popa. Other significant poets are Tone Pavček (b. 1928), Katejan Kovič (b. 1931), Veno Taufer (b. 1933), and Tomaž Šalamun. In some respects these poets are the most experimental now writing in Yugoslavia. While Croatian and Serbian poets are often torn between their native tradition and idiom and the influence of contemporary foreign poetry, these Slovenian poets seem to have completely committed themselves to discovering a modern style accessible to the twentieth-century individual. To consider, further, that some of these poets, Šalamun at their helm, are also some of the most talented now, is to entertain hope for a bright future for Yugoslav poetry.

JURE KAŠTELAN (b. 1919)

Kaštelan earned his doctorate in Slavic studies at the University of Zagreb, and saw action during the war as a Partisan. He has taught Yugoslav literature and literary theory at Zagreb, and has also lectured on Croatian literature at the Sorbonne. Kaštelan's first attempts at poetry go back to student days; he made his debut during the war and is looked upon in Yugoslavia as a foremost humanist and poetic revolutionary, inspired not last by the subtle lyricism of Lorca's folk poetry. His most important collections are: *Biti ili ne* (To Be or Not, 1955), *Malo kamena i puno snova* (A Few Stones and Plenty of Dreams, 1957), and *Izbor pjesama* (Selected Poems, 1964). His 1961 collection of stories and articles, *Čudo i smrt* (Miracle and Death), displays Kaštelan's skill in other genres. He is also a fine translator from several languages.

PARTING

Have you become grass or a cloud that fades.
All the same.

Even along cliffs eagles follow you
even in waters and among stars.

Eyes cannot be separated,
well-springs looking towards the same sea.
There is no parting.

There is no death.

If I listen to the wind
    I hear your voice.
If I look at death
    I hear your song.

(Peter Kastmiler)

BLOOD AND THE STORM

I sliced my veins on sleep's sharp glass.
In my blood the sky mirrors itself.
Here in the Balkans, by the Sava river, by the Adriatic
even the ravens are fed-up with flesh.

Mirrors of terror, show us a picture
without a rope around a neck.
Blood, blood, my blood shrieks
through the earth of my people.

I opened my veins and the bloodstained
seeds of my words.
The hiss of the switch, the storm strikes.
Stones moan.

(Peter Kastmiler)

AS IF IT RESEMBLED YOUR HAIR

As if it resembled your hair
the days on the nape of the stones,
on the soundless unknown wound
the terror flames,

on the youth that whines, on the dawns
that undress like girls into army overcoats,
wearing the heavy helmets of death,
in nameless numbers

the days are a necklace
of steel horrors,
beneath the jaw, beneath the carved face,
hoofbeats threaten.

(Peter Kastmiler)

YOU SHINE IN DARKNESS

You shine in darkness, you dream under shelter
all of fear, all of kindness.
You grow in mountains, swell in dawns,
you regenerate through our lives.

You are above harm, made of dreams,
made of blood, made of flesh,
reveal yourself, spread your wings
above the vultures, above the heavens.

(Peter Kastmiler)

CORPSE IN THE HOUSE

The dead, they do not come from cemeteries.
They live in houses. In mirrors.
They look at themselves in the water's reflection. The mute.
They read, flip pages of books, wear our clothes. Always
unseen. Hungry. Always troubled. Sleepless. Always
blind. Dirty. Confused. Fatigued. Always
transient. Tearful. Noiseless. They steal our bread.
Strip us. Spy on us. And remain silent, silent, silent,
the concealed. Outside of us. Never within us.
Ever present. The unalive—
they live. We the living are in passing.
The guests of death.

(Peter Kastmiler)

VESNA PARUN (b. 1922)

The foremost living Croatian woman poet, Parun was born on the island of Zlarin and has studied at Zagreb. She has been a freelance writer all during the postwar period. Her most important collections of poetry are *Bila sam djecak* (I Was a Boy, 1962), *Vjetar Trakije* (The Wind of Thrace, 1964), and *Ukleti dazd* (Accursed Rain, 1969). She has also written critical essays, plays, poetry for children, and has translated Bulgarian poetry. Her travels in Bulgaria and Romania have proven to be valuable sources for her writing.

YOU WHOSE HANDS ARE MORE INNOCENT

You whose hands are more innocent than mine
and you who are as wise as detachment.
You can read his forehead
and know his solitudes better than I,
and you who remove the slow shadows
of indecision from his face
as the spring wind removes
the shadows of clouds floating above the hills.

If your embrace encourages his heart
and your thighs abate pain,
if your name eases
his thoughts, and your throat
shades his bedside,
and the night of your voice is the orchard
still untouched by storms.

Then stay beside him
and be more pious than those
who have loved him before you.
Fear the echo that encroaches
the harmless beds of love.
And be gentle to his dream
beneath the unseen mountain
on the rim of the sea that roars.

Stroll on his strand. Let the sad
dolphins come to meet you.
Roam in his forest. The friendly lizards

will not harm you.
And the thirsty snakes that I tamed
will be humble before you.

Let the birds that I have warmed
during nights of sharp frost sing for you.
Let the boy that I protected
along deserted roads caress you.
Let the flowers that I watered
with my tears be fragrant for you.

I do not await the best years
of his manhood. His fecundity
I will never receive between my breasts
that have been ravaged by the glances
of herdsmen at fairs
and lecherous thieves.

I will never lead his children
by the hand. And the stories
I so long ago prepared for them,
perhaps I will tell them tearfully
to the poor little bears
left behind in the black forest.

You whose hands are more innocent than mine,
be gentle to his dream
that has remained so unaffected.
But allow me to see
his face, before the unknown years
descend on him.
And give me news of him now and again,
so that I will not have to ask strangers
who wonder at my boldness, and
neighbors who pity my persistence.

You whose hands are more innocent than mine
stay by his bedside
and be gentle to his dream.

(Peter Kastmiler)

OPEN DOOR

And the birds, when in autumn they abandon
their bogs, they whisper good-bye
to the bent and muted reeds
from which the gleam of summer disappears.

And tree separates itself from tree
in the cramped and gloomy spaces
the absent branches lean
one over the other and listen to the river.

And he opened the door to the cabin,
looked at the sky and ran off,
leaving a burning lamp behind
near the unmoving books, leaving the silence
to suffer with the swaying shadows.

O light, transform him into a black stone
at the crossroads under the sharp peaks
from where the wolf's cry descends to the sea.
Let the leaves fall all around him
concealing with their rustling noises
his miserable loneliness.
Let the moon bypass him
not gilding his edges
with the light that soothes the valleys.

Transform, O night, his heart into a flower
on the high slopes of indifference,
where my tears can no longer reach
mixed with the salt of algae.

Gale winds from north and south,
let them upset his capes
surrounding him with hillocks of waves.
And everywhere around him
let the sorrow of the sea spill forth
and the despair, and the eternal absorption.

Then I will close the door
and extinguish the tired lamplight.
And the night will be peaceful,
it will not remember.

He will become nonexistent. The distances
will spread themselves between us
like a good hand of friendship
that screens out the wasteland of the world.

(Peter Kastmiler)

SLEEPING YOUTH

Prostrate on the beach of the shadowy bay
he lies like a fenced-in vineyard,
isolated and turned towards the surf.
His face is smooth and serious.
Around him plays the midday wind.
I don't know if a pomegranate tree
full of singing birds is more beautiful than
the curve of his torso, limber as a lizard.

I listen to the low boom of thunder
that drifts inland off the sea.
And hidden in the leaves of an old agave
I watch this young man's throat turn into hawk
and scream towards the sun, the melancholy flight
through the yellow clouds. And from the bronze
of his luxurious belly rises a black blossoming crag
where magnificent nymphs and fairyqueens rest.

The rocky beach rattles from the retreating surf
and the sea turns gray.
Golden shadows swing over the vineyard.
Columns of clouds climb in the distance
and lightning consumes the wooded bay.

I inhale the summer air of the fields,
and allow the aroma of the herbs to intoxicate me.
I look down at my glistening hands
and my thighs that resemble the surf's spume
from where the oil of the olive flows.

Returning my quiet eyes to him
who sleeps submerged under the arch
of slow storms, ancient as the agave,
I imagine myself full of diffused desires.
How many flapping birds
tremble in the overcast ravine
of that body, that with its silence disrupts
the drone of the sea and the desolate grasses.

(Peter Kastmiler)

BALLAD OF THE DECEIVED FLOWERS

Around the time when the honeysuckle
    blossoms on the embankments,
the flock's golden bleating
    invades the quiet green fields.
Children remove their shoes,
    not to trample daisies,
and swallows arrive out of the blueness
    on a warm Sunday.

A spider has spread a white net
    from the end of a fragrant pine.
Who could possibly think of sorrows now,
    of unplastered rooms and the dead?
Children never believe that the earth
    will swallow the body.

A black smoke hangs above the horizon.
    They say the army is coming.
Whose are these reddening fields,
    whose are these windows on the hill?
Bells are ringing everywhere,
    ringing inside violet daisies.

Why has the spider spread a net?
    why is the army advancing?
Well, read the formidable story
    of the flowers, the clouds, brothers!
In the melting snow
    the tracks of the roe are still discernible,
and a somberness flutters

in the free summits of the conifers.
They say the evil spirit even frightens
the moonlight with its bright reflector eyes.

Why don't the boys go home with their stick guns?
Release the captured bumblebees
and run, run.
An evil spirit walks the moonlight,
children are pulling in their kites,
and the army is approaching.

A hundred tiny hammers forge golden tulips.
Nothing warns the blind larva,
but a child has eyes,
Oh, unfortunate child that will see her father
hanged from the white plum tree in their yard.

Yesterday the honeysuckle
blossomed on the embankment,
and today the drummer's fire
annihilates all of spring.
The bedlam rings above the peaceful clearing,
it rings among alarmed flowers.

Shellfire confused the squirrel,
and the children
fled to the boats moored to the bank,
but the guards wouldn't let them.
Crawl quickly into small anthills,
blow out the candle.

Hidden charges devastated the fishing.
Nothing remains under the sun.
Only so many dried-up graves
and the anonymous ashes taken up by the wind.
Corpses poison the day,
what is man to do?

Daisies folded their eyes from fear,
for the infantry never removes its boots.
A lamb had lost its milk
and remained stunned in the road,
enemy man had come,
to war against unarmed flowers.

(Peter Kastmiler)

359

LET'S GET THE COACHMAN DRUNK

Stop for a moment, driver
of my remaining days of splendor
along the fast river!
Summer's color fades
and a cold settles over the tips of rocks.

Don't scold my empty hands.
They are cautious as gravediggers
who save old songs
for a stone table under the linden tree
where silence is celebrated.

It's useless for you to light fires on the hill
or to call my flocks
scattered across the hillsides
from fear of his brutal horn:
he, whom I had loved so much more than you.

You cannot call them back.
Leave them to wander
the trampled meadows.
Don't reprimand my lost days,
my years, the grieving sisters
of your discouraged youth.

Tonight one should be luxurious
as a garden of sunflowers. Vibrant
and alluring as rain on lake waters
to excite these trees
where the wings of exhausted birds
frighten the fleeting sun.

You must demand a lot from me
so I can accustom myself to your desires
as to forests and to clouds.
You must tell me simple stories
so that your voice can comfort my eyes,
so that I may overcome this need to look back
at the unfamiliar past.

Look! Haven't we already reached
the bend where the hasty water's
murmur becomes anxious?

Get on across! Summer's color fades
and a shadow spreads over the tips of rocks.

This crossroad
will never return.
This is the wedding
of spruce and lake.

Let's go love!

Let's get the coachman drunk who drives our days,
so that he won't know which way our road ascends,
nor in front of which chasm, they will not stop,
these wild horses of love.

(Peter Kastmiler)

SLAVKO MIHALIĆ (b. 1928)

Born in Karlovac, Croatia, Mihalić has worked in Zagreb as a journalist,
and editor, and a publisher; he has also held the post of Secretary of both
the Croatian and the Yugoslav Writers' Unions. Between 1954 and 1968 he
published nine books of poetry, the most important among them *Prognana
balada* (Banished Ballad, 1965) and *Posljedna večera* (The Last Supper,
1968). Mihalić's work is noted for its spareness, understatement, and lyrical
fluency.

METAMORPHOSIS

I would like to know from where
this emptiness comes, so that
I could transform myself into a transparent lake,
the bottom of which you can see, but without fish.

Without shells, crabs, without
the various underwater specimens that at least
carry some kind of name, today I am
without name. Already a little of me is disappearing.

And so speaking of emptiness, I stir
the water in the sea,

sand and other particles churn up from
the bottom. I am being clouded.

I walk through streets with my head lowered as if I were
another lake, dark to everything, afterwards
even poisonous; and we will not talk of
those repulsive little creatures that crawl beneath
    my surface,
so that now I stink, even to myself.

(Peter Kastmiler)

APPROACHING STORM

Look at those clouds Vera why are you so silent
For god's sake I'm not an animal here comes the rain
How quickly it's grown cold
It's a long way back to town

You know I'll never forget what you've given me Vera
Yes we're one now, so what else can we talk about
Yellow clouds usually bring hail
Everything is already still—the crickets and wheatfields

If you want we can even stay
I'm afraid for you for me it doesn't matter
Lightning is dangerous in open fields
We're at our height now (and so damn lonely)

Tonight farmers will be cursing the spilled grains and the
    hard rain
I can't admit that I rely so much on changes
Don't cry Vera those are only nerves
Sensing the storm

I tell you life is simpler in all things
Here come the first drops soon the storm will begin
Do up your dress look even the flowers are closing
I couldn't forgive myself if anything happened to you

Of course this place will remain sacred in my memory
Please walk faster Vera and stop looking back.

(Peter Kastmiler)

AUTUMN

Wagons, if they still ride the country roads, go empty
This summer the plowmen drank too much sun
And now they can't sleep
They hitch their horses, rush out, and then remember that
    the season has ended.

That's what you do to me, you turn your love off, and then
    you don't allow me to forget
With the fury of winds that rip at the last leaves
I follow paths that entreat me to abandon them
Full of shame for their black foreheads, their helpless hands.

Tomorrow (hours before daybreak)
Wagon parts will lie scattered on the barn floor
Horses without legs will raise their heads as if calling

The drunk plowman will expire in his bed

I am already holding my pen, the last candle of my seclusion
And with my verses I lower myself into dreams.

(Peter Kastmiler)

THE MAN WHO DECIDED

    Here somewhere, the world splits in two
like orange halves
    On this day when I decide I am two
halves
    One I throw away; the other is quite
enough for me
    And what is it inside me that frightfully curses
when a decision is irrevocable

    Two worlds, what an enormous burden on these
weak shoulders
    I barely manage for myself; divided world,
you're the chill in my veins
    I don't doubt that I've never done anything
foolish

Who knows the taste of defeat better than
the victor

Everything that is required of me is more
than I am willing to take upon myself
    The real miracles occurred in my fingers
        THE MAN WHO DECIDED, why does this sound
like funeral music
    In my terror I am a branch
preparing to bud.

(Peter Kastmiler)

DRINKING BOUT BENEATH THE OPEN SKY

How drunk I got tonight
That was some drinking bout beneath the open sky
On the banks of the river where the fearful
    dare not go

The darkness swarmed with magicians
They pushed their way through me with cloak-like
    flutters
Long, long, while I was already choking

The river teemed with upside-down fish
    and drowned fishermen

Absolutely, no one bothered with me
And the meteors fell into the world beyond
I must have been very small
    I must have been very good

With my bottle of brandy in my short pants
With my slender hands with my trimmed hair
With my large eyes which needed nothing.

(Peter Kastmiler)

364

MAESTRO, EXTINGUISH THE CANDLE

Maestro, extinguish the candle, serious times have come.
Preferably count stars in the night, inhale your youth.
Your unlistened-to words might bite through the leash.

Plant onions in the garden, chop wood, clean out the attic.
It's better that no one sees your eyes so full of wonder.
That's how your craft is: you can't suppress anything.

You're unable to endure and one night you take up the pen again,
Maestro, be sensible, don't concern yourself with prophecy.
Try to write the names of the stars instead.

The times are serious, nothing is forgiven anyone.
Only clowns know how you manage to pull through:
They cry when they want to smile and they smile
when sorrow demolishes their faces.

(Peter Kastmiler)

SLAVKO JANEVSKI (b. 1920)

Macedonian poet and novelist, born in Skopje. Janevski is regarded as one of the most eminent writers of his generation, in effect a founder, both of contemporary poetry and of the novel in the literature of his republic. He began writing poetry during World War II, and published his first collection, *Krvava niza* (A Bloody Garland), in 1945. Since then he has published six more collections, among them books with such telling titles as *Egejska barutna bajka* (The Aegean Gunpowder Fairy Tale, 1950) and *Evangelije po Itar Pejo* (The Gospel According to Itar Pejo, 1966). Janevski is also a prolific novelist, author of essays, screenplays, and travelogues, and a translator of the works of several Russian and Yugoslav poets into Macedonian.

THE SONG OF THE ETERNAL SAILOR

I left along the distant roads, my apple tree, and now you bloom alone,
my heart is my helmsman, blind yet seeking blue bays,
if I hear the wind in the evening, I forbode your ruin...

Has someone's hunger pulled you out by the roots
as I roamed alone?

When the blackbird whistles shrilly three times at dawn,
do not wait for the sun. Listen, I am still digging roads,
on a mast I carry a black flag from tavern to tavern
and hide the pain under my skull.
Oh does the blue lightning bring you a blue downfall
and do the rains lash you?

I have no more strength to come calm and tall
and to lean my forehead against the sleepy water,
from the blows to rest my hands on the rye until dawn,
and then to go nowhere...
My apple tree, the autumn is already here, there is no shore to sail to.
And so I dream of a secluded, small, and deserted harbor.

(Vasa D. Mihailovich)

BLAŽE KONESKI (b. 1921)

Born at Nenregovo, near Prilep, Macedonia, Koneski studied at both
Belgrade and Sofia. Well known as a scholar as well as a writer, Koneski is
a professor of Macedonian language at the University of Skopje and, since
1950, has acted as editor-in-chief of the scholarly journal *Makedonski
jazik*. He is past president of the Macedonian Academy of Arts and
Sciences. His poetry has appeared in a number of collections; the most
recent of them, *Zapisi* (Marginalia, 1974), reveals the intimate nature
lyricism in evidence also in the one example of Koneski's work that we can
offer.

THE WHEAT

Bold erectness of the girls
who are not afraid of their breasts.
Dark blue flowers are in their hair,
arms firmly pressed to the body,
but the fingers in thoughts thirstily extended
for the first touch.
Ah, that is the girls' choir—

leaning over the border of the stage,
we are waiting for it to hug us like a sweet wave,
but instead we are engulfed in the sounds of a powerful song
that enraptures and then immediately saddens us.

(Vasa D. Mihailovich)

ACO ŠOPOV (b. 1923)

Šopov is from southeastern Macedonia. Following action during the war
as a Partisan and studies at Skopje, he has been active in various editorial
capacities. He currently holds an executive post in publishing. He is author
of a number of poetry collections, the most recent of them *Gledač vo
pepelto* (Gazer into the Ashes, 1970). Šopov, who is also a translator, is
regarded, along with Janevski and Koneski, as one of the founders of
contemporary Macedonian poetry.

POEM

Be the earth unyielding and hard.
Be a mute statue on the table.
Mock the flight of time. And ugly,
grow out of the stone of sorrow.

Be the sky. Be my blueness. A sheaf of stars.
A living time between four walls.
Pierce deep into the heart. Like a spear.
And support me like a caryatid.

(Vasa D. Mihailovich)

MATEJA MATEVSKI (b. 1929)

Matevski was born in Istanbul, of Macedonian parents. He belongs to the
second generation of postwar Macedonian poets. Matevski has translated
French poets, by whom he has been influenced. His books of verse are
Doždovi (Rains, 1956) and *Ramnodenica* (Equinox, 1963).

BELLS

Somewhere bells are ringing. Somewhere far away bells
  are ringing.
The sounds are waves upon the wind whipped along the grass.

Somewhere bells are ringing. Sounding long, clear and tender.
But everywhere the land is still, only rhythmic splashing
against the shore of iron.

Somewhere bells are ringing. Swinging me above the high
  abyss
Running through the cage of sounds hollow and hopeless.

Somewhere bells are ringing. I am a child again ringing
and screaming.
All is closed. Bewitched
I hang myself upon the sounds.

Somewhere bells are ringing. Come strike me! Oh, how
  brave and tame am I
Time strike too, my memory, rudely without end.
Somewhere bells are ringing. Now and long ago.

Everything is painful, underneath the sky. On this grass of
familiar sounds lay me down.

(Dragan Milivojević)

From RAINS

1.
Fear

Arrive sluggish, arrive weary horses of space,
the far-off ominous rumble of forgotten speech,
beating ceaselessly, alone, outside closed windows,
absently, with dull hooves without horseshoes,
on the slippery, on the greasy, on the peaceful earth,
beat, mingling it all darkly.

Where before this bulk, before this cotton
horizon without shape,
before this flesh of earth and night
deeply stirred and dense at the same time,
a flood upon the eyes and the spaces of death.

Where, o where, you and I, noisy and endless sea,
you monotonous
meadow horizontally tired,
longing for the vertical clarity of the wind,
where, o where,
you thick dough of rain and earth
that are to a man, that are to me
stone in the hand and mud in the eyes?

3.
Horses

Arrive sluggish, arrive weary horses of space,
(the rains pale, mute and pathless)
before the cribs of my hands on the window.

Feed, I say, feed yourself sweating horses
damp with the tepid steam that drenches
from the hips of the night.

Neigh wildly, make me shout too,
alight bird with forgotten wings,
weary mare, goat-legged dancer,
let us leap through this window
together and out of it again
and always without stopping to rest,
over the shadowy clarity of space.

(Charles Simic)

GANE TODOROVSKI (b. 1929)

Born in Skopje, Todorovski has edited periodicals and worked in publishing. He is currently Secretary of the Union of Macedonian Writers. Also a literary scholar and a translator, Todorovski published his important poetry collections between the early fifties and the late sixties. *Spokoen čekor* (A Peaceful Step), the title poem of which we print here, appeared in 1956.

A PEACEFUL STEP

1.

Along grey crowded outskirts of the town
dusk is pouring its dark dust,
a messenger of the coming night,
stringing the evening sights
pressing on the senses with a fresh smell of darkness.

I believe: he thought night to be a small square
oppressing the lonesome without mercy
and that is why he went to her threshold
he went into the noisy dusk sunk deeply in himself.

I believe: the bottom of the night attracts the lonesome,
for them it is a haven hidden from all eyes.
Often in her the last prayers of the day are whispered,
often in her one discovers the absence of people.

I believe: he thought night to be a quiet clearing
in which the weary long for rest.

2.

In the morning we found him in the hospital chapel,
again alone on the concrete floor.
He was again lost to loneliness
away from which he prayed to be saved.

His eyes were open and clear
as the crystal-washed sky on nights in May.
They gazed at us with a greeting and proud-pure

and we wanted to find the answer
to the last step in his flight.
His eyes were like two openings
from which we might unravel
what the October night hid from us.

Perhaps like Walt Whitman
we should have called a wreck
that corpse which the morgue
now brought out before our quivering legs,
this frail, this dear face
whom we knew not long ago as an eighth-grade student
whom obituaries informed:
he died in an accident.

3.

Nobody ever knew why he fled from people
laying his body on the rails at the end of the town.
He left silent as steel in a cold embrace.
Hoping perhaps not to remain lost in lonesomeness.

Today his story is remembered.
But the writing in stone will never tell
the silent desperation of this end
the peaceful step he made
on the night of October the eighth.

4.

When reading Sandburg,
I stopped not without reason on that verse
in which the old man says:

"DEATH IS A PEACEFUL STEP
INTO SOFT, CLEAR MIDNIGHT"

(Dragan Milivojević)

VASKO POPA (1922–1991)

A native of the Vojvodina region, Vasko Popa studied at Belgrade, Bucharest, and Vienna, and has for many years worked in publishing. Regarded as one of the commanding figures in contemporary European poetry, Popa is the recipient of a number of awards, among them the Austrian State Prize for Literature (1967). He has to his credit seven volumes of verse. Popa's 1972 volume *Uspravna zemlja* appeared the following year, in translations by Anne Pennington, as *Earth Erect* (Iowa Translation Series, in association with Anvil Press Poetry, London); most recently, Charles Simic has published his second selection of translations from Popa's work, under the title *Homage to the Lame Wolf* (Field Translation Series, 1979).

DREAM OF THE PEBBLE

A hand springs out of the earth
It throws the pebble in the air

Where is the pebble
It didn't return to earth
Nor did it climb to heaven

What happened to the pebble
Did the heights devour it
Did it change into bird

Here is the pebble
It remained stubborn in itself
Neither on earth nor in heaven

It listens to itself
Among the worlds a world

(Charles Simic)

ADVENTURE OF THE PEBBLE

Fed up with the circle
The perfect circle around itself
It came to a stop

Its burden is heavy
The burden within
It dropped it

The stone is hard
The stone it's made of
It left it

So narrow where it lives
In its own body
It stepped out of it

It has hid from itself
Hid in its own shadow

(Charles Simic)

THE CRAFTSMEN OF THE LITTLE BOX

Don't open the little box
Heaven's hat will fall out of her

Don't close her for any reason
She'll bite the trouser-leg of eternity

Don't drop her on the earth
The sun's eggs will break inside her

Don't throw her in the air
Earth's bones will break inside her

Don't hold her in your hands
The dough of the stars will go sour inside her

What are you doing for god's sake
Don't let her get out of your sight

(Charles Simic)

THE VICTIMS OF THE LITTLE BOX

Not even in a dream
Should you have anything to do
With the little box

If you saw her full of stars once
You'd wake up
Without heart or soul in your chest

If you slid your tongue
Into her keyhole once
You'd wake up with a hole in your forehead

If you ground her to bits once
Between your teeth
You'd get up with a square head

If you ever saw her empty
You'd wake up
With a belly full of mice and nails

If in a dream you had anything to do
With the little box
You'd be better off never waking up

(Charles Simic)

EYES OF A WOLF

Before they christened me
They gave me in the interim
The name of a brother suckled by a shewolf

As long as she lives my grandmother
Will call me Little Wolf
In her linen-like Walachian tongue

On the sly she would feed me
Raw meat so I would grow up
To lead the pack some day

374

I believed
My eyes would start to glow
In the dark

My eyes don't glow
Perhaps because the real night
Hasn't yet begun to fall

(Charles Simic)

THE OTHER WORLD

My grandmother puts cupcakes
With lit candles on floating planks

Whispers to them messages
For the dead men and women of our blood
And sends them down the river Karaš

The planks slide down the black water
The little candles struggle through the dusk
And disappear around the turn of the river

Grandmother announces
That they have happily reached
The other world

I have already been there once
To set traps for birds

I didn't know of course
It was my ancestors I was hunting
Among the blossoming willows

(Charles Simic)

IVAN V. LALIĆ (b. 1931)

Born in Belgrade, Lalić completed legal studies there. For a time he resided at Zagreb, and now works again in the Yugoslav capital. Between 1952 and 1975 Lalić published numerous poetry collections, including a 1969 *Izabrani stihovi* (Selected Poems). Technical accomplishment and sophistication in tone are hallmarks of the style of this leading Serbian poet.

THE KING AND THE SINGER

You have to be the stronger one—
my destiny is to endure this song,
this fire, this clownsuit of poetry,

and peer into your squinting eyes,
into that wrath that boils up
and cools down like some unstable metal
in the gusty yards of your smelters.

So I go round and round you
like rain in the desert,
and we are both enchanted:
You in your strength,
I in my inwardness—
and the game goes on through every song—

Every song in which your appetite
sees roses or the delicate forged gold
of vowels, and I my death.

Because—always—
in the space between two words
an empire awakens, unfamiliar,
suddenly disturbed, explosive
like an anthill struck by a spark.

And so in the end, your Excellency,
we are powerless, both of us:
From the wedding of this song
comes a heaven, a more dangerous earth,
and the multitudes we cannot control.

And every night
the familiar stars
slip farther and farther away.

(C.W. Truesdale)

INSCRIBED IN THE SILVER OF THE SEA

and yet sometimes
we were so close to our homeland—
just the other side
of the thinning air—
     we could hear
swifted winged words and surmise
their sense from the lightning's
rough transcription.

     Terrified
we fell back
into a delusion of time,
this false equilibrium
where only at waking
you remember the essential images:

wild fig and chestnut,
a slow flame in the ruins,

and all that we name memory
gleams on the bright point
of a needle, where taste of fire
and silence mingle in the air
above the quick silver of the sea
between two storms.

(C.W. Truesdale)

BRANKO MILJKOVIĆ (1934-1961)

A major Serbian writer who at the young age of twenty-seven committed suicide. He had made his debut in magazines in 1955, and since 1959 had published five collections of poetry, the last of them entitled *Krv koja svetli* (The Blood that Shines, 1961). He also wrote critical essays and translated from the French and Russian. Miljković is a pure lyricist, with a very considerable following among the young poets.

From IN PRAISE OF PLANTS

V

I know your root
But what is the grain from which your shadow grows
Vegetal beauty so long invisible
    in the seed
You found under the earth my headless body
    that dreams a true dream
Stars lined up in a pod
All that is created with song and sunlight
Between my absence and your herbal
    ambitions of night
That make me needed even when I'm absent
Green microphone of my subterranean voice
    weed
Growing out of hell since there is no other sun
    under the earth
O plant where are your angels resembling
    insects
And my blood that weaves oxygen and time

(Charles Simic)

ALEKSANDAR PETROV (b. 1938)

Born in Yugoslavia of Russian parents, who emigrated from Russia after the Revolution. Petrov lives in Belgrade and writes in Serbo-Croatian. He has published twelve books, including several anthologies of Russian, Yugoslav, and Serbian poetry, as well as two collections of his own verse: *Sazdanac* (Self-Maker, 1971) and *Brus* (The Whetstone, 1978). The latter

378

was translated into Spanish and published by Escandalar in New York (*La Muela*, 1979). He spent two years in the United States as a poet-in-residence with the International Writing Program, Iowa City (1972-1974), and is currently a visiting professor, teaching Yugoslav and Russian poetry, at the Ohio State University.

POETRY IN A GLASS CUBE

Poetry and I set out to visit
A man who wants to see me
And poetry so I put on my tie
And let poetry fool around
The man looks at it I suppose he's bored
He lives in a glass cube
And poetry is wild to do its stuff
But I keep it in a safe place
My head in case something goes wrong
Nothing will happen to the glass cube and the man
Can listen but won't just listen
He wants to buy poetry but not just poetry
He wants to buy it along with my head
Wants to hear it in my head
To hear my head echo
Wants to keep it in the center of the cube
Wants to listen wants to watch
Well poetry
Do you want him to make a glass cube
In my head which will echo
Better do you want him to make you
Little glass shoes so that my head will echo
Best or do you want to stop
Poetry what will you do

(Krinka Vidakovic Petrov and Mark Strand)

POETRY IN AN EMPTY DRESS

Poetry in an empty dress

A girl
A sphere
A square
A world

She is the Red Square
She melts red wax
On the square

She is a bridge
Goldengate of San Francisco
Open all hours
Day or night
Promiscuous gateway   brilliant lovers
Endless lovers passing through

She is the marketplace
The Green Wreath of Belgrade
Smuggling nature into town
Advocating the equality
Of convolutions and corns
Stimulating the revolution
of red blood cells

Something like
High tension
Frantic grass a red pamphlet
Rum in a fresh strawberry

Dreams mere dreams

In fact an empty dress
Pants taken off
Clothes without owners
Worn but cleaned and pressed

She can be useful
On various occasions
Even the most solemn
Depending on what you want

Especially if you use the couch
Or are being used in any way

(Krinka Vidakovic Petrov and Mark Strand)

POETRY IN THE UNDERGROUND PASSAGE

Poetry rushes into
The Underground passage at Scales Square
Those who can
Run from it fast
It passes stepping on heads
On Mayakovsky's head
On mine
It shatters Mandelstam's head
Bursts into flame on Dante's
It grows wings on Rimbaud's
It bounces off Eliot's transparent gaze
Like off glass
Then collapses
Breton kisses it
Pasternak baptizes it
Neruda plucks a hair from its ear
And thrusts it into his heart
Hlyebnikov opens its mouth
Bites off its tongue
That flies out sharp as an axe
Marina Tzvetaeva gives it her neck
Lorca his forehead
Nastasievich turns the axe into a flute
Mouths open tongues fly out
Coil into a ball
Saliva sprays all over
The underground passage jumps on my shoulders
Runs out to Scales Square
Soars into the sky
Dives down like a sparrow hawk
And hits an electrical head

Short circuit continuous discharge

(Krinka Vidakovic Petrov and Mark Strand)

381

POETRY VISITS AN OLD LADY

I have enjoyed your company
The white-haired old lady tells me

We were invited to dine at her house
Where she lives with a black prostitute
Tied to a stake
With a clown from the blue period
I noticed that much from my place at the table
And thirty-two masks and poetry
From the pink period

Oh no I said the pleasure is mine
Not to mention the gratitude
If everything goes well
One day you'll have us printed and bound
An expression of thanks

If that's what the old lady wants
What is it the old lady wants

There's nothing we can do
We can't warm her legs
We can only say something in her honor
Something without shame
You have eyes and can see the world
You have hair and in your hair a flower
If that's not enough we can sing a song
You wouldn't understand
But if you prefer we can say
Blue moon yellow horse
Heavenly knees upon my shoulders
Beautiful distance be patient
I'm heading right into you

How nice says the old lady but tell me
How do I shop in a supermarket
I've never been in one before
I hear my granddaughter wherever she is
Aha she was seen in a supermarket

Well the old lady buys tomatoes in the supermarket
Six California tomatoes
Macedonian tomatoes used to be 29 cents but have gone up to 59

The old lady wears blue jeans
Her behind shows through the holes
The old lady is not an old lady
The old lady is her granddaughter

(Krinka Vidakovic Petrov and Mark Strand)

HUSEIN TAHMIŠČIĆ (b. 1931)

Born the son of a Muslim official, Tahmiščić experienced early rebellion
against his bourgeois family; he ran away from home and at age 13 joined
Tito's partisans. After the war Tahmiščić attended technical college to
study electronics, and ultimately the university to study philology. His first
volume of poetry appeared in 1955; to date he has published seven volumes
of poetry, several collections of essays, and a book on Sarajevo (where he
lives).

LIMBO

All round us
Lies the promised land
Regions not yet touched by the sun
Amidst confused landscapes

Wind comes
Wind unsilenced edging
Past locked windows
Past bolted doors
Past tumbling gardens
And doubled-up coiled-up people
Floating down hollow streets
Through the mists of the buried city
Amidst confused landscapes

Between presence and obliteration
In this no-man's-world
Your whirligig is awake
Like a sudden spurt of flame
Like a handful of warm ash
Under the glow of the midnight sun
Amidst rotting landscapes

All round you
Lies the promised end of the world
Amidst the picturesque animated landscape
Of exploding confusion

(Ewald Osers)

LETTER TO A FRIEND

Watch what you are saying
Watch your words
The rest is easily guarded
We're up against a night that shrinks from nothing
Its sentries are alert

You have a fine head
But the thoughts you carry in it are too obvious
Vigilant sentries will easily spot them
We're dealing with axes sharpened
For fine heads with obvious thoughts

Don't ask where and how
The night in which I write to you
Has put away its reasons

Time has moved on
Midnight is drawing near
Sharpened black axes float in a white light
Night opens its gates to them

And that is all
I wanted to tell you
Fine watchful head
Excuse my letter's brevity

I really am
dog-tired

(Ewald Osers)

LANDSCAPES

Slender glow of minaret and silent muezzins
In the sudden echo. Summer sears the early evening
And mothers arrive to turn to stone
Before our eyes, good down-to-earth women
    In our daily world.

We call out, we ring out—in a gigantic bursting,
And they advance down twilight paths, extending their hands
(Like half-burnt candles) across the dying agony
Of the last sunlit blade of grass, good herb-healers they
    In our daily world.

Does the believer's footfall still ring out
As old as the blood reversing the flow of time
    In ancestral sacrifice?

Slender minaret like a lance of sunlight
Snapped in our chest. Night falls on a world turned to stone,
And our destiny (like bitter ice) melts coolly
On the muezzins' tongues, cunningly lulled to sleep
    In our daily world.

Broken, dismembered, we're threaded on sneers—
Yet they stare through the children into dusk-filled ravines: silent.
Silent and pathless the land in the night that is falling
For those without support, for the clouded mind, for the seed flung
    In our daily world.

Why have we come to the very bottom of the landscape
Of gilded bells swinging aloft
    Stubbornly silent?

(Ewald Osers)

STONE AND ASHES

Night is divided. And behind us:
Shadows unevenly growing, our permanent loss.
Nothing
That could serve as an answer for the demented snows

(Or for the nights that keep warm underneath them
And which we pollute).
Nothing
Is all that's between us
Except the earth in our pupils and the wind between our ribs,
And the water thickening and stiffening all the time.
Unfamiliar our thought in the translucent blackness of the ice's crust:
Phosphorus flaring from the bones of our scattered skeletons,
Spirit encouraging disbelief in our crucifixions.
And a new sound ringing through the void,
Out of tune with the quiet of daybreak or nightfall: at the base of the horizon.
Instead of the body, bronze shall groan, set swinging
By the weight of our silence.

(Ewald Osers)

RED MOON HIGH IN THE SKY

The hyenas are coming
The hyenas are laughing in the night
                    Dane Zajc

Again the spiked signal
Of pursuit has gone up
Ominous shades are lurking all round us
Fiery tongues reaching out for us
Red moon high in the sky

Out of the night
Come the hyenas
At our spell-bound heads
In the night

The hyenas are laughing
At our bleached gnawed skeletons
Red moon high in the sky

Shine brightly
Red moon high in the sky
Show us the road down which we've not yet fled
Red moon high in the sky

(Ewald Osers)

EDVARD KOCBEK (1904–1981)

The ranking elder of contemporary Slovenian letters, Kocbek studied at several institutions of higher learning, including the University of Paris. A short story writer as well as an essayist, Kocbek has kept his poetic work spare and lean; the center of gravity of his published collections of poetry lies between the mid-thirties and the late sixties. In his insistence on craftsmanship and spirited contemporaneity Kocbek has been seen as a spiritual confrere of young writers like Šalamun.

THE STICK

What shall I do with my stick
now that it has begun to outdistance me?
Shall I throw it on the fire of a shepherd,
or give it to the lame man on the road,
or to scouts reconnoitering the promised land?
Or shall I raise it in the air
to still the people's tumult,
or use it to trip my brother
so he breaks his leg in the dark?
Or shall I throw it into the sea
to save a drowning man,
or plant it in a field
to stand in the wind as a scarecrow?
Or shall I hang it in a pilgrim church
to increase the number of relics,
or bury it in a wood
so the bailiffs can't find it?
Or shall I give it to an ignorant father
so he can use it to tame his son,
or leave it out in the dew
so it turns green again?
Or shall I hand it to a choirmaster
to harmonize the voices,
or give it to an eager boy
to use it to prop up his tent?
Or shall I divine a spring with it
in order to water the desert,
or use it to conjure bread
from a stage magician's hat?
No, I will do nothing of the sort,

for all that is risky and foolish—
I will break it over my knee
and throw it down a deep ravine,
so that its heavy notches
may measure my fall.

(Veno Taufer and Michael Scammell)

DIALECTICS

The builder demolishes houses,
and doctor advances death
and the chief of the fire brigade
is the arsonists' secret leader,
clever dialectics say so
and the Bible says something similar:
He who is highest shall be lowest
and he who is last shall be first.

There's a loaded rifle at the neighbor's
a microphone under the bed
and the daughter is an informer.
The neighbor goes down with a stroke,
the microphone's current fails,
and the daughter goes to confession.
Everyone clings to a ram's belly
when sneaking from the cyclops' cave.

I hear in the night discordant music
coming from the circus tent,
sleepwalkers walk the highwire,
wobbling with uncertain arms,
and their friends yell underneath
to rouse them from sleep,
for whoever is up must come down
and whoever's asleep, let him sleep more soundly.

(Veno Taufer and Michael Scammell)

GREGOR STRNIŠA (b. 1930)

Born in Ljubljana, Strniša studied German literature at the university there. Also a playwright, he lives as a freelance writer. Of his numerous books of verse the best-known are *Samorog* (Unicorn, 1966) and *Brobdingnag* (1968). The translations here reprinted show an articulate, craftsmanlike poet recognized as one of the best now writing in Slovenian.

A SHIP

On a long, long trail, a dark ship.
On the masts the silvery jewelry of the moon,
southern winds leaning against the large sails,
showers pour along the grey path of the sea.

The sailors' eyes burn like coal from thirst.
The fiery captain seeks the way in darkness.
The sun hewn from heavy gold
and the seven stars of Orion shine above the ship.

High up on the masts the flags
flutter gaily like apparitions.
With the stern wrapped up in sea grass,
the depth glides like a cloud, like a star
over the creatures that are rigid.

(Vasa D. Mihailovich)

THE PRIESTS

When we burn out the heart,
not a single drop flows.
When we burn out the whole heart,
a small black mask remains.

The mask has the likeness of God
or resembles the devil.
More often than God's face
we see the devil's.

A river flows down the snowy mountain.
In the cave a mute flame gutters.
With its help we have changed man
into a shriveled black effigy of the devil.

(Michael Scammell)

THE GRAVES

Through little lands two horses drew
a clay effigy of the dead king.
Clay horses beneath a clay moon.
The little years passed quickly.

A clay maiden in the cart
played fast songs on a silent zither.
Now and then a long wave rocked the ship
far out to sea of the departing potter.

Ship of oak foreseen in a green acorn,
apparition of light on the sword's tip—
tiny country made out of clay
in the big black country of earth.

(Michael Scammell)

TOMAŽ ŠALAMUN (b. 1941)

Although born in Zagreb, Šalamun is one of the foremost Slovenian
writers. His several books of poetry, ranging from *Poker* (1966) to *Amerika*
(1972), have already earned him wide recognition, and his poems have been
translated into several languages. Our examples of Šalamun's verse show
him to be a whimsical poetic personality, and at the same time a craftsman
of true aim and depth.

HOMAGE TO A HAT & UNCLE GUIDO & ELIOT

Just as Clay became a world champion
because there was something wrong with his leg
I'll be a great poet
because they double-crossed me
with Frank's blue cap
sent for Christmas 1946
and since then I've left him out of prayers
song of songs of pansalamunian religion
terribly democratic people's institution
which takes in everything
from stamps biscuits Tzilka, Horak, Parmesan
to that poor idiot
who drank his hotel away in Ventimiglia
and faded out somewhere in the world
just as our prayers fade out
its last important reformer was uncle Guido
known among the folks
for his invention of a new pipe for a steamboiler
but that was not his main occupation
his main thing was
watering flowers
just like Spinoza

a bit taller though
meditating on and off on death
buying us ice cream
each day made new
that was between
magnolia Brandenburg & America

two days ago Eliot died
my teacher

(Tomaž Šalamun and Anselm Hollo)

391

JONAH

how does the sun set?
like snow
what color is the sea?
large
Jonah are you salty?
I'm salty
Jonah are you a flag?
I'm a flag
the fireflies rest now

what are stones like?
green
how do little dogs play?
like flowers
Jonah are you a fish?
I'm a fish
Jonah are you a sea urchin?
I'm a sea urchin
listen to the flow

Jonah is the roe running through the woods
Jonah is the mountain breathing
Jonah is all the houses
have you ever heard such a rainbow?
what is the dew like?
are you asleep?

(the Author, Elliot B. Anderson, and Charles Simic)

BULGARIAN

INTRODUCTION TO BULGARIAN POETRY

By Alexander Shurbanov

The first poems that have come down to us in the Bulgarian language are contemporaneous with the Christianization of the country and the introduction of the Cyrillic script in the ninth century. They are strongly colored by the attitude, the diction, and imagery of the new religion, but they are also clearly secular: Bulgarians, like other Slavs, are developing a culture beside those of other Europeans, and expressing themselves in language that everyone can understand. The first Bulgarian poets we hear of are not churchmen or mystics but popular tribunes. Their subjects are neither otherworldly nor personal; for the most part, they speak of the survival and the good of the nation, ideals that become the purpose of their work as well as of their lives. This singlemindedness will become a feature of Bulgarian poetry throughout the troubled history of the country.

If we look at the mediaeval annals of Bulgaria, it will not be difficult to understand this peculiarity. The young ethnic state, one of the first of its kind in Europe, was faced from its inception with the formidable challenge of taking root and growing on the edge of the Byzantine Empire, on territories formerly possessed by it, and in a cultural context it determined. For seven long centuries Byzantium was both respected teacher and feared enemy, an ambivalence that made of Bulgaria a wary pupil that knew its own mind.

Then came the Ottoman conquest. An exquisite Balkan civilization that was flowering into a fourteenth-century renaissance provided a rather poor defense against the brutal military expansion by the Turks. Almost overnight Bulgaria, its social and cultural leadership obliterated, was reduced to a darkened province of the northwestern Ottoman Empire. Its Christian society was ruthlessly and methodically suppressed by its new Muslim overlords. Together with their neighbors Bulgarians became part of the laboring peasant mass, and the light of their national heritage seemed to have been extinguished forever.

Yet it was among those illiterate peasants that the spirit of poetry endured, and, with the burning of the books, it returned almost spontaneously. Five hundred years of endless foreign occupation yielded a tradition of folk poetry whose variousness and artistic riches are splendid. The spiritual consciousness of people was preserved in and by these anonymous songs through the misery of subjugation, rapine, and slavery. And it was from that popular tradition that modern Bulgarian poetry took its beginnings during the period of the national revival that culminated in the liberation of the country in the last quarter of the nineteenth century.

Deprived of cultural continuity, the new poetry of the Bulgarians took some lessons from other, more fortunate, European literatures. It is not hard to see in Bulgarian poetry of modern times the beneficial influence of the Russian and French romantic and symbolist schools, as well as of later international trends and movements. Underneath all these, however, there runs a current of rhythms, themes, sentiments and images that are essentially derivative of native oral tradition, and have flowed easily into modern poems from folk song. And, together with these, the modern poets have inherited some typical traits of folk singers: self-effacement and absorption in communal aspirations, a striving to express the communal experience in ways that can touch the most contemporary sensibility. Bulgarian poetry has by and large remained resistant to aestheticism and its lesser relation, snobbish sophistication, despite the receptivity of the poetry to foreign influences and experiments. And its own sophistication has seldom resulted in obscurity. As one of the first national poets put it in the ninth century: "I prefer saying five words / my brothers and sisters can understand / to thousands they do not."

The poems presented here are, we believe, a fair sampling of the best work in recent Bulgarian poetry. They are by a group of important writers belonging to quite different generations, yet all are equally active on the contemporary scene. Each has contributed in his or her way to the creation of that sense of continuity in change that makes Bulgarian poetry what it is today—as old and as young as the nation it comes from.

ELISAVETA BAGRYANA (1893–1991)

Elisaveta Bagryana is the most respected and honored of Bulgarian poets, a recipient of several national and international awards. One of the poems offered here, "The Only One," was published by Alain Bosquet in his anthology of one hundred of the world's best poems. Bagryana was born in Sofia and brought up in Sliven and Turnovo, two colorful cities at the foot of the Balkan range that have left an enduring mark on her work. She has played a central role in the literature of her country ever since the publication of her first remarkable volume of poems, *The Eternal and Holy,* in 1927. Her writing is admired for its insights into the feminine, and for a music that has its source in folklore. Some of Bagryana's early poems have remained unequaled both in her own work and in that of others; yet her recent works offer examples of similar beauty, if less passionate and more reflexive in tone.

THE OLD FORTRESS

Once more I walk the round of Hisar's stronghold
in this forward spring.
The park, roused, lifts its shoulders
and the gilded domes
of flowering shrubs
glow like a cathedral.
Even ruins a thousand years old
are paradisiacal
with birdsong.

What atavistic calling through the centuries
takes me unaware?
I shiver with the thought that birds by the thousands
nesting
in this ruin
of mortar and stone
as though in some realm of their own
could be the spirits
of unknown thousands on thousands
of Thracian thralls
who built this fortress
with inhuman toil, with their last efforts—
for Romans, the foreigner
and master.

Today these spirit-birds fly in and out
in great clouds,
making their nests,
raising children and their descendants—
and they grow up
and sing,
the very fortress singing in a key of green
through them now.

Nature and time are unafraid
of contrasts never to be reconciled,
and blend them in a harmony
impossible for us.

(Jascha Kessler and Aleksandar Shurbanov)

RIVER RUN

A river strangely named—
Lielupe—
runs, runs
through my closed eyes,
through my memories,
through my reveries,
through my dreaming
when the slimechoked,
meaningless day is done.

Mornings, the boat departing
calls me
with its basso hooting
from the quay.

And there I am, sailing upstream
once more
into the wind
amongst the strangers on board.

Ahead, distant
settlements approach,
drifting on a different journey,
two never dreamt-of shores.
The rushes stoop,
the water lilies flare
like white smiles,
and all of it passes, running away.

Illusory! Not all of it—
we're what's passing,
while shores, trees, stones
remain.

May two white words of ours
reach their haven,
like lilies
on the shoreside waters
of someone's memories.

(Jascha Kessler and Aleksandar Shurbanov)

[The Lielupe River runs through Lithuania. Bagryana visited there some years ago.]

THE ONLY ONE

Was it you yesterday?
Is it you now?
Will it be you with me tomorrow?

That face I see
against my closed eyelids,
that silhouette with a changing shadow
walking with me always,
that voice at morning
wakening me, making me sing,
that name I call you by—
are they yours? are they yours?

Is it you or is it
the image and name
of my thirst,
that waits trembling
like the thirst of the fruited earth
for a rainbearing cloud?

Is it you or is it
the image and name
of my grieving
for that one,
eternal,
faithful companion—
as the moon is to earth.

Is it you?

(Jascha Kessler and Aleksandar Shurbanov)

BLAGA DIMITROVA (b. 1922)

Blaga Dimitrova was born in the northern Bulgarian town of Byala Slatina, but grew up in Turnovo. She is the author of some of the most impassioned love poems in the language. Her collections include *Time Reversed,* 1966; *Condemned to Love: Poems about Vietnam,* 1967, and *Selected Poems,* 1968. Her recent books have grown increasingly meditative and dramatic, drawing problems of our world into the sphere of personal experience. Dimitrova's philosophical prose, unmistakably that of a poet, has also been popular, and has been published in various translations abroad. Her novel, *Journey to Oneself,* appeared in London in 1969, shortly after its first Bulgarian publication.

IF

When you return,
if you return,
it's only then you'll find you're gone.

The streets will lead
here there everywhere,
and only your somewhere will be standing still.

Your greeting will be
mumbled, downcast,
and it will find a stranger's welcome.

Guiltily you'll walk into your home,
looking all about
as though it's a house forgotten in some dream.

And you'll run your fingers
over your self missing
among the books and things all rearranged.

And you'll know
something has been rearranged,
not merely your house but the world as well.

Just like that and naturally—
so as to occupy
the space taken up by you.

(Jascha Kessler and Aleksandar Shurbanov)

A WOMAN PREGNANT

She walks, slow and solemn,
gravid with the sorrows of the world.
Amidst that roaring she alone can hear
what's inside of her,
that echo beating there.
A tiny fist is pounding there,
trying to punch through the strictured dark
and leap out into the sun.
She passes through ruins
and through the felled forests of the killed.
It's all booming in her heart!
Every crater aches inside her.
She gathers up the knife-edged shards
of a disintegrated world
in her swollen womb—
to make it whole and new once more.
She walks, wise and anxious,
containing in herself as in a chalice
that frail laughter of the future.
And carries the globe of the world ahead.
                    [Vietnam, 1967]

(Jascha Kessler and Aleksandar Shurbanov)

CHILD AND SEA

The child tears itself free of my hands
to stay by itself with the sea.
On the edge of the earth it stands,
at a threshold that crumbles into eternity.
A little voice sprouting from the sand
says to the hundred voices of the sea,
—Come and catch me if you can!
And the white-maned element is there,
its endlessness abandoned for a game.
A wave dashes at his heels,
reaching for him hissing, splashing,
and then retreats, pretending fear,
driven off by little naked feet.
"The sea's running away!"
                    Though next moment the wave

returns once more, acting as though enraged,
threatening him with its foaming mouth.
He runs from it shouting boldly,
slapped by a wet palm from behind,
both of them braying laughter.

                                        So on and on
child and sea will play forever,
unless I part them
forcibly, intending to carry
the child away to hands and words,
none of which he comprehends,
leading him imperceptibly,
and step by step, towards exile
from his rapport with the world.

(Jascha Kessler and Aleksandar Shurbanov)

INTRODUCTION TO THE BEYOND

You were dying, fully conscious,
having summoned the incredible strength
you needed to go calmly,
not crying, not moaning, not shuddering—
to keep fear away from me.

Your hand went softly
cold in my hand,
and led me gradually
beyond to death,
so as to introduce me.

Once you would hold
my little hand as gently,
and lead me through the world
and show me life—
to make me unafraid of anything.

I will follow you
with a child's trust
towards the silent land,
where you went first
to make me known there.

And I shall not be afraid.

(Jascha Kessler and Aleksandar Shurbanov)

[From a suite of poems entitled *A Requiem*. These were composed for Dimitrova's father.]

STEADY

Hesitating's my one steadiness.
Infinite possibilities,
infinitely variable.
No matter what I choose to take,
it will be my mistake.

Any choice makes me poorer.
So as not to blame myself,
I'd prefer that chance
do my choosing for me.

But chance never
lifts a finger
to take responsibility away.
Feedom's what it offers me.

Whatever I may reach for,
all the rest is lost to me.
If I grab for it all,
I'm sure to catch nothing.

Choosing's a sort of little
death I'm always putting off.
Let me hesitate, I say.
Tomorrow, yes, but not today!

The infinite's alive yet,
really all's still possible.
When my hand snatches at that one thing,
I'll obliterate it forever.

Till then everything's still mine!

(Jascha Kessler and Aleksandar Shurbanov)

BOZHIDAR BOZHILOV (b. 1923)

With more than thirty volumes of poetry behind him, Bozhidar Bozhilov is easily the most prolific of living Bulgarian poets. He was born on the Black Sea coast in the romantic city of Varna and while still in his teens engaged in underground political activity. His first book of poems came out in 1939 under the title *Third Class*. Bozhilov has travelled widely, always on the lookout for new sources of excitement. At the same time he has managed to turn a good deal of his everyday life into somewhat rambling, quietly lyrical, and often charming verse. An interesting collection of poetic impressions, *American Pages*, reflecting his recent stay in the United States, has been translated and published in this country. Bozhilov is editor of the literary periodical *Pulse*.

ETERNITY
   for William Meredith

1.
Man is immortal.
Man's surely not his body?
Man is the thoughts
this body carries.
And the husk of flesh
preserves the salt of our thoughts.
Yes, the body's the plastic spool,
which goes on unwinding and winding,
carrying the tape of thoughts.
In myself I have,
almost as though they were my own,
the thoughts and feelings
of my father and mother.
I've already conveyed
my own thoughts and feelings
into my daughters' hearts.

I am immortal,
because the body's the skin of the soul,
which is sloughed off like a snake's skin,
so that the flesh of thoughts
may live forever,
the deep river of dreams
carrying memories and reveries,

and gathering my feelings like a mirror
into the stars' eternal drift.

The mortal animal dies,
it sheds its skin
so that the body
of immortal thoughts is left.
And with my body's shell
I pass into nothingness,
but my brain's a seed,
elsewhere planted
and already growing,
even setting the fruit
of its next ripening.

I'm the fragile cup
for the liquor
(that time brings from far away
and takes farther off yet),
the drink with millions of curative herbs
and a cry of the enormous brain
of the people
and of mankind in me.

I am eternal
born.

2.
A fleeting, chance meeting.
Diffident smiles.
Some words.
Two hurried snapshots on the terrace
below the dome of the Capitol in the sky,
accidentally spoiled.

But the snapshot-word of an image that flashed
identically
in two different poetic systems,
and in two hemispheres of the globe,
and its two different philosophies of happiness,
can never be spoiled
by distance or time,
by love or hatred,

because this image entered the words
of two languages,
fusing them in the four-dimensional
that till now has only been seen by him,
who sits calmly on the green lawn down there
in his bronze deathlessly-dying body
on a chair of bronze,
embellished by sparrows
with eternity.

(Jascha Kessler and Aleksandar Shurbanov)

THE WISE ONES

Ten wise men standing stock-still
in an airport waiting room.
I make eleven.
We're silent, frozen as yogis.
Without so much as the blink of an eye.
We're looking over the sea,
which we can't get to,
which is none of our affair,
which looks like a painted illusion.

Planes taking off and landing.
Engines roaring and fading.
But that's none of our affair.
Each waits for his far-off hour,
which may never arrive.

Sometimes someone hears his name.
Or the name of the city he's flying to.
Or his flight number.
Then he picks himself up and disappears.
Though it goes unnoticed.
Because a new eleventh one has come,
imperceptibly, invisibly.

We must always be eleven.
Individuality's irrelevant.
Irrelevant the directions,
the hours,
months, seasons of the year.

We wait.
Stock-still, like stony idols.
Wise and eternal.
Others are rushing and anxious.
Glancing at their watches,
studying schedules,
asking for information,
leafing through newspapers,
drinking coffee,
buying souvenirs,
talking into telephones.

We stay silent, stock-still.
Always eleven.
When the twelfth arrives,
if ever he does arrive,
he'll be called Peter
and the cock will crow.

(Jascha Kessler and Aleksandar Shurbanov)

LILIES

Some day when I leave,
still kidding as usual,
even about life and death,
then all your sorrowful words,
whose thorns pricked me,
will flower like the lilies
of the Annunciation.

(Jascha Kessler and Aleksandar Shurbanov)

TROUBLES

It was an awful day.
Troubles piled up on me,
as though I'd gone through
an earthquake.
Yet in this day of misery,

so awful for me,
my pangs deliver a poem that's happy.

(Jascha Kessler and Aleksandar Shurbanov)

KENTISH SESTINAS

The Kentish landscape all about
outside the window calmly hears
absurd discussion of a drought
while drizzling rain makes trickling tears
and the translucent pond without
reflects some July-ripened pears.

Peace and contentment, come to me!
With English wisdom fill my breast!
Let fear be an infrequent guest
and let me find to joy the key!
A rainy morning let me be
without a rhyme, without a jest.

What every song needs is a special scenery
across which, like a bird, it can soar higher,
to settle briefly on some distant wire,
a secret sentinel amidst the greenery,
wisely and bravely pondering the machinery
of sunsets with a coin of golden fire.

It's nine P.M. and still the night
is lingering in some distant meads:
the half-dusk, not yet superseded,
still trembles on the rain-damp site,
expectant still of love's delight—
a love that is no longer needed.

Upon this field beyond that ancient tree
a strange unearthly craft should be descending,
and lips not of this world and eyes distending

408

should ask: What kind of starry ecstasy
can you expect, what joyous victory,
from these green apples, tasteless and offending?

The tongue of Shakespeare and of Eliot:
a distant horn call beckoning from covert?
Each word of yours is a new rhyme discovered,
like flashing sabers' thrust and cut
engaged in mortal duel but
to glorify some wanton lover.

England remembers countless ancient kings:
in countless churches brood their sculptured faces.
Round ancient towers ancient glory paces,
from wind-blown trump a battle signal rings—
but then the wind, that huge bird, folds its wings
upon the plaques at poets' resting places.

What is modernity? A concrete highway
and bridge have slashed across an ancient wood.
And yet the new is like a toy that could
one day be just a memory: a byway,
a monstrous stream of motorists, a skyway
beneath a canopy of ancient cloud.

The Kentish rain: yes, that is real rain.
It comes down slowly, clearly not intending
to stop for hours yet, rain never-ending,
without a single flash of sky again.
It seems to me like an old angry man
jealously amidst dripping bushes standing.

(Ewald Osers)

GEORGI DJAGAROV (b. 1924)

Georgi Djagarov was born in the village of Byala, south of the Balkan Range, and fought as a Partisan during the Second World War. Since then he has been involved in political life, both as poet and as a citizen. At present Djagarov is Deputy Chairman of the State Council, the highest executive body of the government. His first collection of poems, *My Songs*, was published in 1954, and his infrequent subsequent publications have been consistently greeted with admiration. His passionate nature comes across with equal power in both his intimate and his "public" poems. *The Prosecutor*, a topical play written by him in the Sixties about the Cult of Personality in Bulgaria, enjoyed international success, and its English edition was prefaced by C.P. Snow.

AUTUMN

I said goodbye.
If that's the way you want it to be,
all right . . .
And the sea's raging in all its might,
this Black Sea.

Dearest,
we used to meet here
by chance
like wandering gulls.
And stars,
multitudes of stars,
poured over our evenings . . .
But the world's faithless,
look—the twin tracks on the shore
are gone,
altogether lost.

And last words perish in nothingness,
unuttered,
as before.

Autumn,
autumn,
this way she passed,
a slandering of yellow leaves she tossed.

I'm caught.
Condemned.
I know—
I'm to die
at love's stake.

The wind blows,
dragging brushwood to my feet.
Like a hangman the horizon
drops its hood.
Someone's sobbing.
Why is someone sobbing?
It's useless sobbing.
I dissolve in twilight.

I burn.
I'm going up in smoke,
disappearing
like the heretic to death condemned,
disappearing like a boat
in the violent gulf,
disappearing in darkness like the land.

I said goodbye.
If that's the way you want it to be,
all right . . .
And the sea's raging in all its might,
this Black Sea.

(Jascha Kessler and Aleksandar Shurbanov)

NIGHT

Why are you not asleep, child?
Why do you weep?
What grieves your slumber?
The leaves that drop rattling in the dark?
Or that trolley's far-off clanging?
Those voices fading out?
The moon
climbing over the distant hills?
Or the owl screeching in the wood?

411

Or that streaking star
come to disturb the stillness
of this trouble-wearied world?
Or my wakefulness
long before the new day's here?

Why are you not asleep?
Sleep, child, sleep.

(Jascha Kessler and Aleksandar Shurbanov)

LYUBOMIR LEVCHEV (b. 1935)

Lyubomir Levchev was born in the beautiful mountain town of Troyan.
When he published his first book of poems, *The Stars are Mine* (1956), he
was recognized as an extraordinary talent writing in the tradition of
Bulgarian revolutionary poetry. His contagiously romantic vision, his
youthful ardor, and his often striking imagery have won Levchev an
audience chiefly among the younger generation. The books of this poet
have been translated and published in more foreign editions than those of
any other Bulgarian poet. One of them, *The Mysterious Man,* came out in
the United States in 1980. Levchev was recently elected President of the
Union of Bulgarian Writers, a position usually held by the most prominent
figures in literature.

CAPRICCIO 9

I
was descending
the mountain
when I saw them
descending from the sky.

Astonishingly beautiful.
Dazzling white with blood-red meridians.
Their silky domes wavered
over new fields
            new railway stations
            over new worlds
            and new...

"Say, what a lot of parachutes!"
the old woodcutter exclaimed.
"War's not on, is it?!"

But I knew they were your temples,
my fantastic Twentieth Century.
Invisible
up to the domes themselves.
Made from pure thoughts,
from limpid liberty,
from lunatic logic,
from equations
and revelations...

And in those temples
you—Twentieth Century—instructed us
never to pray no matter what
to anyone.

"They came down just like quail!"
the old woodcutter exclaimed.

And I saw shrivelled domes.
And I saw vanishing temples.
And I saw myself
grounded...

Twentieth Century, you're leaving!
The take-off.
The sky-dive—

another twenty years are left.

(Jascha Kessler and Aleksandar Shurbanov)

CARMEN, OR THE MADE-UP GIRL

Once,
in the pearly once-upon-a-time,
before holidays,
                    that shadowy once-upon-a-time—
we used to turn into green giants.

413

We used to down brandy boilermakers.
We used to smoke long, goldtipped cigarettes.
And tell about our love affairs . . .

That was when I made up
                    the made-up girl.

My turn came,
and I had nothing
as yet to divulge,
so—
I made up a tale that we'd kissed
behind the transformer.
And that we'd whistle our secret tune:
"Tell me, Carmen, that you love me . . . "

Naturally they all laughed at me.

But when I was alone
and small
in the late hours
composing obscure verses,
someone started whistling urgently below my window:
"Tell me, Carmen, that you love me . . . "

I opened up
flickering like flame
and saw you
                    —oh loveliness—
passionate,
hair in the moon streaming,
all lit up by the midnight wind,
and with the whispering of iris petals:
"Here I am—
just as you made me up!
Call me Carmen!
And believe that I love you! . . . "

Ah, dearest Carmen!
How often
you've come to comfort me,
urge me on toward victory,
or lull me
                    through my misery . . .

By axioms betrayed.
By comrades-in-arms betrayed.
And by myself betrayed...
By you alone—
                    the made-up one—

never yet betrayed.
When I lost the way,
you stayed with me.
When I was out of ammunition,
you stayed with me.
And you asked for nothing more
than a boy whistling:
"Tell me, Carmen, that you love me..."

Darling, stay with me through this night too!
Ruffle my hair,
already going gray!
Go on kissing me!
Draw my fever away!
Because I'm burning down to ash,
and the wind is strewing me
through time.
And now I know
(what some day everyone will know)
that actually
the made-up one was me.

(Jascha Kessler and Aleksandar Shurbanov)

APPASSIONATO

I love the sky at night,
for it alone is naked.
And the light of day keeps me
from seeing the universe's nakedness.

Ah, that starry nakedness!...

Don't get dressed!
I want to see you.
To decipher your matter,
which draws me so strongly.

Love me.
Be my night.

(Jascha Kessler and Aleksandar Shurbanov)

SECRET LOVE

Were you in love with me?
Or just lying to me?

I wander through vacant alleys like a blind one.
Scurrying meaninglessly.
Hunting that corner—

that secret niche for the absurd explanation.

But right there—
in that blankwalled pit—
I'm caught,
                    surrounded
by spectral poses
by strange shapes
by dismal plasters
and frozen marbles—

Gradually
I catch on.
This is the yard where failed statues are stowed.
Rejected from shows.
Left unrecognized.
By their own sculptors abandoned...
So
where can they go?

Swaddled in nylon
or dust,
or naked
and turned to the wall
                    they continue
playing through
their doomed parts.

They're silent, these rejects.
I'm silent also.
And time's silent.
I hear a voice though.

A galvanized trash bin, all battered
deluding himself
that he's an art work
because he's here,
and out of his yawning lid
filling the darkened town with his prediction:

Brothers!
The day is coming
when they'll recognize us!
They'll beg our forgiveness,
        brothers!

Ah, maybe the day will come
for everything in this world!
It will! . . .

Except for you, my love,
you're never coming back.
I understand—
                you've spelled out
                what's incomprehensible to me.

But I—
I lack the talent of marble.
I lack the toughness of zinc.
I'm kneeling.
I'm kissing
hope's fading footprints—

Don't reject me!
Don't reject me.

(Jascha Kessler and Aleksandar Shurbanov)

ABANDONED CHURCHES

A heart without love is like
a church without god.
When the neighbors suspect that no one's there
they'll enter
and start pulling it apart—
a slab from the altar,
a piece of tiling,
an icon-lamp...
why, just for souvenirs!
And afterwards the real thieves will come.

A heart without love
experiences joy
even when torn apart.
But how terrible is a dying church!

The best thing then is to spread a rumor
that specters are
haunting
the abandoned church,
stalking at night, knife in hand,
and in daytime swinging
from the chandeliers...

After that expect
heaven to send a flash of lightning,
so that your church burns down
immaculate.

(Ewald Osers)

CONFESSION AND SALUTE TO THE FIRE

Interrogated by
the whips of my own imperfection
I've no strength left...
I confess...

I confess that
poetry is a criminal attempt

to translate
the pulsations of the universe
into the simple mortal language of our hearts.

I confess:
poetry is brazen witchcraft.
Poetry permits
the gods
to assume
human shape
and simultaneously brands humans with divinity.

I confess
that from the very beginning I was aware
that such witchcraft is punishable
by burning at the stake.

And so—
hail, our poetic
life-giving death-bringing Fire!

I kiss your ashes,
brothers and sisters at the stake.
And wholly to you I commit myself,
my red
my warlike love!

(Ewald Osers)

WE STOPPED

We stopped.
Precisely under the sign
with the inscription
"Stopping prohibited!"
and then you told me:
"This is my favorite spot!"

A bridge.
Out of rails and planks.
You rusty rainbow!
And the sign with the inscription:
"Stopping prohibited!"

There is no river running under the bridge.
Railway tracks run down there.
Track signals
beat their wings.
And midnight shuntings thunder.

And when
the engines passed
under our feet
and the steam clouds closed around us
totally invisible
and totally unique as we were
we kissed.

There must be millions of freight-cars
shipping our kisses now
to all four corners
of the earth.
We left
when I told you:
"This is my favorite spot!"

A bridge
A miraculous one.
An unforgettable one.
Between childhood and maturity.
And a sign with the inscription:
"Stopping prohibited!"

(Atanas Slavov)

LAST NIGHT

Last night I was coming back home alone.
Somebody tapped me on the shoulder
as good friends do sometimes.
I turned around cheered up
but there was no one there.
Only on top of my shoulder glowed
the ruddy palm
of a single
huge
heavy
chestnut leaf...

... I was jolly all evening.
So my mother started an argument
looking at me from under a heavy eyebrow:
"You've met old friends again, eh!... "

(Atanas Slavov)

LUCHEZAR ELENKOV (b. 1936)

Luchezar Elenkov represents some recent trends in contemporary
Bulgarian poetry. Though he was born in 1936, only a year after Levchev,
he published most of his books in the seventies. Their titles alone reveal the
interests of the artist in the space age: *Perpendicular Towns, Forbidden
Grass, Saved Space, Zero Kilometer, Continental Craters.* Unexpected
associations and striking metaphors, a weighted conciseness of expression,
and intellectual daring are the distinctive features of his style. Even when he
writes of the past, Elenkov sees its tragedy and heroism from the viewpoint
of a contemporary who is preoccupied with the world's future. At present,
he is co-editor of the international literary journal *Druzhba,* and Secretary
of the Bulgarian Writers' Union.

STONE

You lay in the
catapults. And took aim
at the slitted embrasures.

But no one could see you.

You were a millstone.
And sweated for our daily bread
with us.

But no one could see you.

They made a wall with you.
And you broke the waves
before they reached
the tranquil haven.

But no one could see you.

Beneath the visionary chisel
you became the echo of flesh.
Everyone admires
your cold conquest now.

(Jascha Kessler and Aleksandar Shurbanov)

SAUNA

Superheated steam
floats above a planked roof.
A train stop it might seem
in a dank, ashtree wood.

But a door bangs open
wide as the world.
Out dashes a woman—
and heads for the freezing well.

Bare thighs ringing,
and distant ringing laughter.
A brief sun gliding
over the rounded shoulders.

Something that resembles
paired doves below the curved neck.
And vapor that trembles
wrapped about them yet.

Running feet that vanish
beyond the wall of a weir.
Then the watery splash
and a swan soaring free there.

(Jascha Kessler and Aleksandar Shurbanov)

POET

Foliage, bees and flowers . . . walking alone under low skies.
A bug vanishes out of the air down the pigeon's gaping bill.
Without work, life and thought, you could scarcely
make out even a way forward. Ask the twilight,

422

ask the ones who know you—how long will you last?
A city of names is raised up on posters like obits for those yet alive.
More and more of them. And you barely glance at the next one's work.

A swelling emotion. The world's almost gone from your mind. And then—
a tulip seems a sort of bell. And you blare out the news.
Which the neighbors scarcely hear. Go on without them,
to where the dew is fuming over black loam. And in spring or fall
they dig wells, potatoes, turning up punctured skulls.
Start there. Measure yourself against the gods. You'll grow tall
in that other land—the sky above these sprigs.

(Jascha Kessler and Aleksandar Shurbanov)

TEN GIRLS

Enraptured by earth, rain and sunrise,
restless and beautiful as cornflowers
I met them on the road. And behind me exploded
the laughter of ten girls.

Like a volcano my heart came to rest again
and I could feel it dividing into ten Luchezar Elenkovs
and each called out behind me:
                              "Don't go away. We are here."

And they waved their arms like windmill sails.
I hurried to catch up with our company.
But they poured over me their ten-voiced laughter:
                              "Love us . . ."

And so they receded on the road like a rose-colored cloud,
the ten girls I'd invented in my mind.

(Ewald Osers)

AMPHORA

Amphora
      as an emotion

Through the blue fig-trees
as the suns come out
those young Greek women.

Amphora
      as a metaphor

Under southern sails
it evokes that country
where the slaves artistically
created art.

Amphora
      as a form:

Precise and severe
as a denial
      of God.

(Ewald Osers)

NINO NIKOLOV (b. 1933)

A prolific younger Bulgarian poet, Nikolov was recently writer-in-residence at the International Writing Program, The University of Iowa. In the same year (1979) his *Poemy* (Selected Poems) were published. Widely traveled, Nikolov often records poetic experience of places abroad, as in some of the poems printed below.

THE WINDOW

The woman from the balcony opposite
undresses when the sun comes out
and sits in the wicker chair
with her eyes shut.

The rain
falls on the traces of past rain,
and sun and clouds alternate.
I like it when she undresses,
the woman from the balcony opposite:
the sunlight plays about her slender neck,
her breasts rise like two rising suns.
When I get up in the morning I ask myself
what will the day be like—because I know
the woman undresses for the sun.

Ewald Osers)

CONFUSION

When confusion reigns
and the waitress begins to think
that the hungry exist for her,
and the air hostess begins to think
that the passengers exist for her,
and the pharmacist begins to think
that the patient exists for her,
and the reviewer begins to think
that the author exists for her—
and not the other way about:

When confusion reigns
I too, my love, begin to think
that you were made for me.

(Ewald Osers)

DON'T MOVE AWAY

As into the green cool of the sea
under pines' shadows on the beach
I slowly enter into your world.
Such is the fever of emotions
with which you carry me away
to crystal depths.

You string your words together
like islands scattered long ago
in this sea which engulfs us.
Are you speaking or is your voice tracing
a world that's free and beautiful,
a happy atlas?
You carry time in your eyes.
Why should we hurry?
Two sunrises pour out their light over me.
My hands now are two oars, at rest
far from the cliffs on the shore.
A wave is rocking me—
accidentally, just like this meeting,
and perhaps too late,
because my eyes are already afraid of the horizon.
Don't move away, whispers my fear
but I am silent. With my silence I do not name
that which at any moment can stand
between us . . .

(Helsinki)

(Ewald Osers)

WATER OUTSIDE THE WINDOW

Water doubles this world.
Whatever it takes it returns twofold.
The trees grow with their roots in the sky,
the clouds crawl over the ground
and the roses shed their petals on them.
Depth exchanges height.
The fault, of course, lies with the water
when it is crystal-clear.

(Karlovy Vary)

(Ewald Osers)

YIDDISH

INTRODUCTION TO YIDDISH POETRY

"I'll find my self-belief in a dustpuff of wonder," Jacob Glatstein's persona sings, looking up at "a salvaged half-star / that managed not to be killed." The sense of the miracle of survival, of life itself, shines through that grateful utterance. How often we read, in introductions to anthologies of writing from third-world and otherwise troubled countries, that despite all odds their peoples have succeeded in preserving their ethnic and cultural identity. If this holds true for anyone it certainly does for the European Jews, who out of untold strata of diaspora and persecution have emerged with a splendid body of contemporary poetry. As a supranational phenomenon this work deserves to take its place alongside the recognizably "national" poetries that this volume features.

Writing by Jewish poets is supranational and national at once. On the one hand this anthology includes poems by poets of Jewish background from several countries (e.g., Porumbacu, from Romania), some of whom may insist that background is incidental to their present loyalties and interests. On the other hand it would be difficult to find poets more Jewish in their poetic identity than Paul Celan or Karl Shapiro, whose languages are German and English, respectively. With the international community of poets who write in Yiddish we are on safest grounds in insisting that language both wells from and determines outlook, poetic vision, and sense of spiritual belonging. How generous the vision, and how truly universal the sense of spiritual presence, we can let the poems themselves document. Certainly some of the poems display overt concern with the classical themes of Jewish religion and culture: the Bible, folklore and mysticism, the holocaust, life in the *shtetl* or in Israel. Others portray the joy and sorrow, faith and doubt attendant simply upon being human. We can cherish, for example, J.L. Teller's exuberant "Minor Key" for the sheer poetic puzzle that it presents. Who is speaking? Even such a reminiscent poem as Sutskever's "A Cartload of Shoes" could imaginably resonate with overtones of a Christian who has also suffered. (Although it would have had to be someone of the quality of János Pilinszky.) It is only when all the facts are in, when we know about background, identity of the author, and all those other factors of which it is the business of properly edited anthologies to inform us, that we can experience the richness, the special flavor and atmosphere, of poetry written originally in Yiddish. In this respect, too, Yiddish poetry joins hands with fellow members of the family of nations and poetries here represented.

Most of the poets whose work appears here were born in northeastern countries—Estonia, Lithuania, Poland, the Ukraine, other areas of the USSR—and have settled in Israel, Canada, and the United States. The voices of Glanz-Leyeles, Glatstein, and Moishe Leib Halpern belong distinctively and energetically to New York. Glanz-Leyeles, for example, writes a tribute to Madison Square (of the famous Garden) that makes us remember Carl Sandburg's Chicago poetry; Glanz-Leyeles addresses that "Manshape, Flatiron Building, proud wonder, / You cut into the Square, a hero-juggler," while Halpern writes movingly of the Sacco-Vanzetti outrage. Others, like Gabriel Preil, also a distinguished poet living and working in New York, do not commit themselves so directly to thematic concerns; in fact, Preil, in "Memory of Autumns at Springtime," goes back to a scene of his childhood and makes it communicate with "an American autumn": "The Lithuanian autumn is no longer a boy / who makes apples dance in his basket: / it has aged like a villager / shouldering an empty bucket." Among those who came to live in the Western Hemisphere, perhaps Rachel Korn gives us best a feeling for subject that lies as far as possible from what surrounds a person; through the sands of the Sahara "Thirty-one camels / make their way / without a leader." There is no gainsaying the argument that this oriental imagery is in some ways close to what Korn herself, in her poem, identifies as "my longing, / the treasure of my sacred hope." That sacred hope is evoked in equally powerful but very different ways by Itzik Manger in "Abraham and Sarah" and by Moshe Yungman in "The Messiah"—the hope and faith of aging Sarah on the one hand, the hope of messianic redemption on the other ("Still . . . on the dark cobblestones / Of the streets two blind old Yemenite grandfathers / Drag heavy bundles of dreams to their Moriahs").

But it is the finely tuned scale of sensitivities that will make this poetry, like any, rewarding in the deepest sense. As elsewhere the translators do, of course, deserve some of the credit for conveying the insights of poetic craftsmanship. In the poems of Alquit-Blum, Segal, and Teller, values of sound come across with remarkable brilliance and delicacy both. Alquit-Blum's poetic instrument resonates with great silence ("The hand of creation / weeps."); In Segal we overhear the sounds both of nature and of human mourning ("Memory sweeps up like a mist / from the roads and fields. / Sweet, the mouth of the fresh rain / in the new summer's sun"). The precisions of sensibility are perhaps clearest in Kulbak and Leib. In Kulbak's "Summer" the delicately erotic assumes queenly importance on levels both thematic and metaphoric: "Today the world unwrapped itself again, / fat bud, full earth, green whispering, / everything trembled the way taut girl flesh / trembles with the sharp joy of becoming a wife. . . . " Leib is a powerful craftsman, and the translations of the three poems that we have render a particularly needed service in preserving the precisions of prosody and stanzaic order as well as of image. Let me quote just the concluding stanza of "The Pyre of My Indian Summer": "And trees—blue wax—in the void's brightness cool / Like erect candles, full of holy dread. / And silence keenly marks the sere leaf's fall, / And keener still the unrest of my tread." It is a complex poem, and the role of metaphor in the first line of this quotation seems inseparable from the role of the eye, with the subtlety of trees as "blue wax" hardening into the clarity of the "erect candles" in the next. The net effect is one of mystery—simultaneous movement and standstill. Powerful feeling is at work in Sutskever's "Toys" and "A Cartload of Shoes"; both work with what Ruth R. Wisse, in the anthology *Voices within the Ark,* has named "the dimension of absence" (New York: Avon, 1980, p. 242). But the most triumphant example of the eye of the poet as painter comes in Sutskever's "The Banks of a River": they "light up silvergreen and violet, / then darken again." Truly, as Susan Sontag said in *Against Interpretation:* "In place of a hermeneutics we need an erotics of art" (New York: Farrar, Straus & Giroux, 1966, p. 14).

ELIEZER ALQUIT-BLUM (1896-1963)

Born in Chelm, Poland, Alquit-Blum emigrated to the United States in 1914, and worked as a tailor in New York. Best known as an expressionist poet and co-editor of the modernist Yiddish literary journal *Inzich* during the thirties. He also wrote for the New York Yiddish *Morning Journal.* His one verse collection, *Two Paths and Other Poems,* appeared in 1931.

THE LIGHT OF THE WORLD

The soundless
light drifts over
the distant
sea.

The white
body of a woman
lies upon the silent
earth,
over the layers of frozen
treasures.

Only
the unspoken

429

word fills itself
on silence.

The hand of creation
weeps.

(Howard Schwartz)

ASYA (GAY) (b. 1930)

Born in Vilna, she worked in a Russian labor camp in 1941, and was
repatriated in Poland. In 1951 she went to Israel with Youth Aliyah, and at
present lives in Jerusalem. Her poems have appeared in a variety of
periodicals, and have been translated into Hebrew. Her book, *Trembling
of Branches,* was published in 1972.

THE DEER

for Jacob Glatstein

I don't know whether the gray deer knows
That his horns can wound,
But he surely knows the sources of clear water.
In his kingdom,
Where the Divine Presence and danger live side by side,
He is not afraid to risk breaking his slim leg
When thirst gives him wings.
And he does not pursue the shepherdess,
Whose pitcher gives drink to the passerby.
His sense of smell, polished by the wind—a sail
Bearing him upon the waves,
And when he raises his head, drenched with water,
His horns dip into the sun.

(Gabriel Preil and Howard Schwartz)

Float up again.
We will descend to our own depths.
Give the seadeck to the preying fish
With their rough scales.
Take your head out of that strange dishevelment.
Put on the blue *yarmulke*—the sky of your city,
Her beams wrapped in psalms.
See how the East flames up,
An eternal candle.

(Gabriel Preil and Howard Schwartz)

A. GLANZ-LEYELES (1889-1966)

Born Aaron Glanz in 1889 in Wloclawek, Poland, Glanz-Leyeles came to the United States in 1909. He has been using the pseudonym A. Leyeles for his poetry. In 1919, in collaboration with Jacob Glatstein and N.B. Minkoff, he founded the In-Zich movement of Yiddish poetry and the journal of the same name. His books of poetry are: *Labyrinth* (1918), *Young Autumn* (1922), *Fabius Lind* (1937), *A Jew at Sea* (1947), *At the Foot of the Mountain* (1957), and *America and I* (1963). He is also the author of two plays and of a collection of essays entitled *World and Word* (1958).

WHITE SWAN

Girl, your young loveliness
is a white swan
Winging above me, casual in triumph.
In swooning oblivion I lie
And it
Quite unaware of what it does
Pecks at my heart
With pink and tender bill.

(Keith Bosley)

MADISON SQUARE

1.

No rooftops to rest on.
Lines soar and are lost on
The heavens they thrust on.
  Severe and male,
The feeling—chaotic,
Cooped up, anti-gothic,
Commercial-quixotic.
  Mere chance for style.
Pent energy, crouching.
Trapezes high-pitching.
From wills overreaching.
  Anxiety-state.
Grandiose in its plainness,
Derisive of meanness.
Diverse in its oneness—
  New York the Great.

2.

Manshape, Flatiron Building, proud wonder,
You cut into the Square, a hero-juggler,
Your hands and feet flung back, head to the fore
For take-off—but manlike you check your desire.
Straight up your line shoots, Metropolitan, the thunder
Of your song thrills every poor man's dream on the Square,
Your mocking laughter lands on asphalt under.
Carved forms crowd round, each one an obelisk,
Heroic towers puncture the solar disk,
And through a gap a corner of the Garden is slender,
Fame of a former generation that more and more
Loses itself in New York and runs from care and risk.
But Fifth Avenue laughs and Broadway flaunts a smile,
With neither will nor time to ask is it worthwhile
To change their form or style.
They change. The Square—a chance halt, a surrender
Whose passion in one stormy moment spent
Bore giant sons of iron and cement.

(Keith Bosley)

432

JACOB GLATSTEIN (1896-1971)

Born in Lublin, Poland, he emigrated to the United States in 1914, where he helped found the introspectivist In-Zich group of Yiddish writers. Along with Abraham Sutskever one of the two towering figures of contemporary Yiddish verse, Glatstein published thirteen volumes of his poetry. His *Selected Poems,* from which the following selection comes, were published in the English translations of Ruth Whitman.

RUTH

Roaming through a field of threshed stubble,
I came upon your solitary tent.
I hestitated, thought of running away,
but was caught and held
by your sky over my alien world.

The night smelled of barley harvest.
Reeling with the scent, I kneeled
and uncovered your feet
with my shaking fingers.

Spread your long skirt over me,
let my toes seek yours:
let this night not be more sacred than you.

You're sturdier than the night,
which wants to defeat us.
Twine my bare legs tightly in yours,
my legs feel numb.

Joy in my head, the sound of my people's
prayer of thanksgiving:
it is good to be your untouched captive.
Dreamlike, I hear your low
subdued Judaic words.
I translate them closely into Moabite
until I detect your music, your very breath.

(Ruth Whitman)

I STILL REMEMBER

I still remember the half-lit meaning
of words, the drowsy meaning
of the alphabet flocked together,
like frightened birds
huddled beak to beak in sleep,
dreaming of angels;
of souls washing themselves
with white rags
in brimming glasses of water
beside covered mirrors;
of words looking for happiness
in houses where melancholy
walks in stocking feet, on tiptoe.

(Ruth Whitman)

LOST

You've lost the kingdom
of eternal childhood.
Not death nor time
nor seas nor deserts
can tarnish the crown
of those young days
that lived in you
and younged as you aged.
The destroyer of your city
destroyed your cradle,
shot your father through the heart
and struck a whole world
of Jews—struck the heart of your youth.
No matter how long you're fated
to grow old,
your years are already beheaded.
Aging, they'll come
to a cave of silence,
walled up
and frozen
in an isolated inconsolable
deleted silence.

(Ruth Whitman)

IF YOU SHOULD BE SURPRISED

(The Bratzlaver Rabbi to his scribe)

If you should be surprised, Nathan,
at the human being today and his doings,
let me tell you that in the Soul Warehouse above
there's been bedlam for a long time.
There isn't a single halfway decent soul left
for reworking,
to make a new beginning.

That whole treasury has been depleted.
Every soul is tattered, stripped, badly soiled.

Oh, years ago souls would turn up
pure and clean as at birth.
It was a pleasure—just smooth one out
and off it went—into a new body.
Now there's not a new soul to be had even if your life
    depended on it.
They're all degraded, secondrate, wornout,
sinned through and through.

So they sit above us—a patch here, a scrap there.
For some generations now, bodies have been wearing
these miserable leftovers.
Haven't you noticed, Nathan,
what a terrible deal
mankind is getting lately?
So if you should be surprised
at man and his works,
what can a poor son-of-a-woman do
with a secondhand soul?

(Ruth Whitman)

MOISHE LEIB HALPERN (1886-1932)

Halpern was born in Galicia, and was a participant in the Czernowitz Conference of 1908, where Yiddish was proclaimed as a national language of the Jewish people. He came to North America in 1908, and lived in New York and in Montreal. Considered one of the leading figures of the avant-garde group Di Yunge, he is especially remembered for his two volumes of verse *New York* (1919) and *The Golden Peacock* (1924).

GO THROW THEM OUT

When people come with big muddy feet
and open your door without a by-your-leave,
and begin to walk around inside your house
like in a whorehouse in a back street—
then it's the heart's finest joke
to take a whip in your hand like a baron
teaching his servant how to say good morning,
and simply drive them all away like dogs!

But what do you do with the whip when people come
with corn-blond hair and heavenly blue eyes,
bursting in like birds briskly flying,
lullabying you as though with lovely dreams,
and meanwhile stealing into your heart,
singing, taking off their tiny shoes,
and, like children paddling in summer brooks,
dabble their pretty feet in your heart's blood?

(Ruth Whitman)

THAT'S OUR LOT

Young fisherboys sing like the endless sea.
Young healthy blacksmiths sing hot as the fire.
Razed buildings on abandoned streets, we sing
Like the emptiness there when it rains.
In parks the children play and sing together.
In their song lives the love of a good mother.
But we, it seems, were not born of a mother.

Misfortune, singing, dropped us in the road.
We sing unlucky songs for no good reason.
Perhaps like parrots swinging inside cages,
Like frogs at dusk between the swamp and grasses,
Or laundry outdoors when the winds are blowing;
Or else like scarecrows, that have been forgotten
In fields, when fall has preyed on everything.

(Kathryn Hellerstein)

RACHEL KORN (1898–1982)

A Galician by birth, Korn lived in Lvov until 1941, when she sought refuge in the Soviet Union. Soon after the war she emigrated to Canada. Since 1928 she has published eight collections of her poetry, the most recent of them being *Bitter Reality* (1977). She lives in Montreal.

THE THIRTY-ONE CAMELS

Thirty-one camels
trudge through the white burning sands
of the Sahara,
laden with the load
of my longing,
the treasure of my sacred hope.

Against them
Sahara stretches her feverish brown body
indolent and weary.
Eagles fly over her with flaming wings
and lions soothe her with lullabies
on starry nights.

Thirty-one camels
trudge among white caravans
of bleached skeletons
with annihilation
in their eyes.

Thirty-one camels
make their way
without a leader,
without a guide.

(Howard Schwartz)

PUT YOUR WORD TO MY LIPS

Put the word to my lips
like a signature at the end of a page.
Send me—where? I don't know—
for who waits there but the dark syllable?

I have been anointed with sadness
like the queen of endless night.
She does not know whether she is in a dream
or someone has imagined her.

Perhaps she has only been gambled away
to Fate's winning hand—wagered
as a stake and forfeited,
abandoned to the wind, to the unknown?

Put the word to my lips
and lead me like a child by the hand
to the border outlined by tears,
frontier at the country of night.

(Seymour Mayne and Rivka Augenfeld)

MOISHE KULBAK (1896-1940)

Moishe Kulbak was born in Smorgon, Lithuania and switched, at an early age, from Hebrew to Yiddish as the language of his poetry. His first book of verse, *Songs,* appeared in 1920. He settled in Byelorussia and became associated with the Minsk group of Soviet Yiddish poets. After early prose works written in an expressionistic style, he wrote a novel, *Zelmenianer* (1931), and a mock-epic, *Childe Harold of Dissen* (1933), in the style of socialist realism. He was arrested in 1937 and died in a labor camp in 1940.

438

Since his rehabilitation in 1956 his books have once more begun to appear in the Soviet Union.

SUMMER

Today the world unwrapped itself again,
fat bud, full earth, green whispering,
everything trembled the way taut girl flesh
trembles with the sharp joy of becoming a wife...

Like a cat I lay in the middle of the field
where it splashed, flashed, sparkled and glistened,
one eye smeared with sun, the other—closed,
and silently rejoiced, silently laughed...
over miles of plain and forest and valley—
here's where I lounge about—a splendid hard steel.

(Ruth Whitman)

MANI LEIB (1883-1953)

Mani Leib (Brahinsky) was born in the Ukraine and emigrated to New York in 1905 after a period of involvement in the Russian revolutionary movement. He was an active contributor to the anthologies of the aestheticist New York group Di Yunge, and published numerous collections of his poetry, including a volume of poems for children.

THE PYRE OF MY INDIAN SUMMER

The pyre of my Indian summer burns
Out now in rings of smoke and drops of gold.
I am dumb and devout as my hand turns
The ash upon the last coal-star gone cold.

And night and villages. The crickets play
On moon-flutes to my soul a mournful tune.
Upon white grass at fences going gray
The pumpkins are as yellow as the moon.

And trees—blue wax—in the void's brightness cool
Like erect candles, full of holy dread.
And silence keenly marks the sere leaf's fall,
And keener still the unrest of my tread.

(Keith Bosley)

IN LITTLE HANDS

In little hands she holds an open book,
Her head is leaning back, becalmed in grief,
The sun sinks at the window, sends a brief
Red glow across the pages of the book.

Where is big sister? She has gone away,
Away—why and for how long no one knows;
At home that evening as she always was,
By night she had for ever gone away.

And afterwards the mother wrung her hands
And with a kerchief covered her old face;
She lit the candles, stuck each in its place,
Wept without tears and wrung and wrung her hands.

And everyone at home was grieving then
And everyone at home was watching mother
And quieter grew their tread about each other
And everybody's talk grew quieter then.

... The book falls from her little hands. She weeps.
Her childish lips are quivering with fear.
"Like my big sister I shall go from here
As soon as I am grown up...." And she weeps.

(Keith Bosley)

PSALMODIST

Let shrieking steel and gray stone be set
For green grass; the clash of hammer and tongs
And the rush of wheels—birdsong;
Grating of saws—the summer cricket; the pious poet—

The merchant. And as in earliest and early days
The grass will shoot its stalk through the rock
In rhythm with the cricket; and over the clanging shock
Of steel—the bird's song will range far and away.

So too the poet: pious, from market noises,
From business bustling and the wheels' clatter
He will move apart with parchment, ink and feather

To the holiness of the word, which in its commonness
Is strong—to stitch prayers for the world: to knit
The heart of God, to unbind the sorrow on silent lips.

(David G. Roskies and Hillel Schwartz)

ITZIK MANGER (1901-1969)

Born in Bukovina, where he acquired strong interest in Yiddish folklore
and in German literature. His first collection of verse appeared in 1929. In
all his work—in a number of volumes of his poetry as well as in plays—he
shows interest in the modern adaptation of Biblical themes and motifs.
After fleeing the Nazis he lived in London and in New York, emigrating in
1967 to Israel, where he died.

RACHEL GOES TO THE WELL FOR WATER

Rachel stands by the mirror and plaits
her long black braids,
she hears her father cough
and wheeze on the stairs.

She runs up to the windowseat:
"Leah! It's father! Quick!"

And Leah comes to the door,
hiding her trashy book.

Her face is drawn and ashen,
her eyes red and weepy.
"Leah, you'll ruin your eyes,
you've read enough today."

And Rachel takes the pitcher
and starts off towards the well—
the twilight is blue and mild,
it makes you want to cry.

As she crosses the dark field
a rabbit flashes by.
Chirik!—a little cricket
chirps in the deep grass.

And in the sky there shimmers
an earring made of gold:
"If only there were two,
I'd like to have them both."

A piper whistles near her:
tri-li, tri-li, tri-li—
The air is full of dusk and hay
from all the cows and sheep.

She runs. It's late. The Good Book says:
a guest waits near the well,
today the cat has washed her face,
today she fasted too.

She runs. And high up sparkles
the earring made of gold:
If only there were two,
she'd like to have them both.

(Ruth Whitman)

AUTUMN

September. The gypsy and the nightingale
don't know what to do with themselves.

A moonwalker sleeps near the cool river,
stripped of all his dreams.

Ophelia, sick Ophelia!

Weariness in a silk dressing gown
smooths the gray valley with her fingers.

She smooths and says incantations and wakes
the blue marvel of once upon a time.

Ophelia, sick Ophelia!

The green eyes of the September night
look wearily through the windowpanes.

In those eyes—cries of wild cats—
what are those eyes to us?

Let's run away. But where can we run?
The blind lantern stands and watches,
and the doors and windows are closed—

Repent, Ophelia!

This marvel took off
from Aladdin's magic blue lantern,

and the gypsy and the nightingale
don't know what to do with themselves.

And Itzik Manger sleeps on the hard ground,
stripped of all his dreams.

(Ruth Whitman)

GABRIEL PREIL (1911–1993)

Born in Estonia, Gabriel Preil has lived in New York since 1922. A bilingual poet, Preil has published seven volumes of his poetry in Hebrew, in addition to publishing a volume of verse in Yiddish. He has been the recipient of numerous literary prizes in the United States and in Israel, including the Jewish Book Council of America Award and the New York University Neumann Award. The poems appearing here are from Mr. Preil's *Selected Poems*, published in English translation.

MEMORY OF AUTUMNS AT SPRINGTIME

The Lithuanian autumn is no longer a boy
who makes apples dance in his basket:
it has aged like a villager
shouldering an empty bucket.

But in the same fall no longer remembered
by even one clock in the world,
I borrowed some evening sadness
and a glass to clarify
the dark at noon's heart.

Once, perhaps, the American autumn also
shouldered an empty bucket
but now, when the colors of apples
sing in the basket, I am held back
from asking anything.

By day and night sorrows block my doors.
And when I cast a glance in the glass of light
something forest-black strikes me
and like the Lithuanian autumn, I am no longer a boy.

(Laya Firestone)

HOW TO READ THE FIRST LINE

The first line of a poem
is a hawk not letting go of its prey
or a forest ablaze with lightning
on all its blind sides.

Later perhaps a strip of meadow will rustle,
a thin horn of moon will appear—
until at an imaginary gate of victory
all praise of existence will arise
and the sun will stand as in Gibeon.

(Laya Firestone)

LINES ON THE RAMBAM

Maimonides never saw the snow.
But he felt the frosty fire of wisdom,
felt the whiteness
which embraces the world with its roar
and becomes a balanced and measured silence.

Maimonides never saw the snow,
but a strong wine of consolation
calmed in his glass.
Through a cloud the banners of mercy
waved. And his eye caught faith
shining out of the woods.

(Linda Zisquit)

YELLOW SUNSET

A yellow sunset splashes in the window
of the yellow-shaded house.
And in the shade of the door a man,
at his feet a small yellow lake.
In his hand a glass
with a yellow-colored drink.

On the back of a wall, as if woven,
the dark yellow shadowhead of a woman emerges
and her long hand draws from an orange
a yellow flame.

The burning sun grows in the center
of a cool yellow world,
surrounded by yellow villages and towns,
and somewhere the wing of a bird shimmering yellow
utters a singular, transparent song.

(Linda Zisquit)

JACOB ISAAC SEGAL (1896-1954)

Segal was born in Podolia, in southwestern Russia, of a family of scholars.
He emigrated to Montreal in 1911, and published his first book of poems in
Yiddish, *From My World,* in 1918. In Montreal he was the editor of two
short-lived literary journals, and wrote for the city's Yiddish newspaper,
The Canadian Eagle. For many years Segal was president of the Canadian
Jewish Writers' Association. His books of poetry include *Poems and
Praises* (1940), *A Jewish Book* (1950), and the posthumous *Songs for
Jewish Children* (1961).

CANDLE

Your innocence snuffed out,
your frail thinness curtsied to Death—
today at the corner store I bought
a bit of wick for just a penny.

Pour a shard full of oil,
dip the white wick:
for the broken twig,
for my daughter's young body.

Memory sweeps up like a mist
from the roads and fields.
Sweet, the mouth of the fresh rain
in the new summer's sun.

My daughter, little bride,
in your white silk and silver veil
we must give you away.
From the night's distances
the young summer's wind
blows against my dark windowpane.

(Seymour Mayne)

REST

Such a calmness
upon your face—
Your every word
is a translucent song,

thin, perfect,
scintillating,
sharp as October's
crystal ring.

A breeze freshly blossoms
over your weariness,
a guest blows in—Death,
clothed in the lightest snow.

Throwing off his hat
and coat, he holds up
his grayish
proud head,

joins us at our table,
crosses his legs—
on a silvery afternoon
talk is filled with light.

(Seymour Mayne)

ABRAHAM SUTSKEVER (b. 1913)

The commanding figure among living Yiddish poets, Sutskever was born near Vilna, and was a leader of the Young Vilna group before the war. He escaped the Nazi occupation of Lithuania by fleeing to Russia, in the process saving considerable archival materials from the Vilna Jewish Museum, of which he was then curator. In Israel, where he lives, Sutskever edits the Yiddish literary magazine *Goldene Keit*.

SONG FOR A DANCE

I invite you, child, to dance. You come. I bow my blond head,
bending it down to the ground.
Eager, warm, this is how I see you:
a yielding ear of wheat, your arm, your knees,
the singing outlines through your skyblue dress
and—your eyes—velvet joy—
I forget where I am. I become a springtime stream,
singing my heart out through every atom
of my blood. And then, and then—and only this is real—
we twine our hands together and we both
are equal in our dance.

And suddenly we go, a stormy journey
through big dark woods, over night and day,
over time. The world hides somewhere in a corner.
I don't know who I am. I think: I'm evening gold, I'm you.
I'm even a swallow. . . . Shouts fly by, rivers, cities,
everything—in our dance, with red and violet
and green. You cry. I listen. Your look falls on me like fire.
I think, you're a sailboat somewhere on the sea
and I a salty wind, in a duel with you.
You struggle. I spank you with foam. And we both
are equal in our dance.

(Ruth Whitman)

THE BANKS OF A RIVER

From a high mountain I see how the banks of a river
shimmer. In the distance
near the horizon they darken and wrangle,
then light up silvergreen and violet,
then darken again. I look down
into the river where my face's tinder is quenched
and my body shines clear, transparent,
and I say to the east, west, north, south:

Look and see
how beneath choked leaves and houses
in cold riverwriting my name is written.

Broadcast it all over the world.
Amen.

(Ruth Whitman)

POETRY

The last dark violet
plum on the tree,
delicate and tender as the pupil of an eye,
blots out in the dewy night
all love, visions, trembling,
and at the morning star the dew
becomes airier—
that's poetry. Touch it without
letting it show the print of your fingers.

(Ruth Whitman)

TOYS

Love your toys, my darling,
your toys even smaller than you.
And at night when the fire is sleeping
cover them up with stars.

Let the little gold horse put his nose
in the cloud shadowed grass so sweet.
Put the little boy's shoes on his feet
whenever the sea eagle blows.

Put the panama on your doll's little head
and a little bell in her hand.
For not even one has a mama,
and they cry to God in their bed.

Love them, your heirs to the crown,
I remember a day of woe—
seven little streets with dolls in each one,
and not one child in the town.

(Seymour Levitan)

A CARTLOAD OF SHOES

The wheels hurry onward, onward,
What do they carry?
They carry a cartload
Of shivering shoes.

The wagon like a canopy
In the evening light;
The shoes—clustered
Like people in a dance.

A wedding, a holiday?
Has something blinded my eyes?
The shoes—I seem
To recognize them.

The heels go tapping
With a clatter and a din,
From our old Vilna streets
They drive us to Berlin.

I should not ask
But something tears at my tongue
Shoes, tell me the truth
Where are they, the feet?

The feet from those boots
With button like dew—
And here, where is the body
And there, where is the bride?

Where is the child
To fill those shoes
Why has the bride
Gone barefoot?

Through the slippers and the boots
I see those my mother used to wear
She kept them for the Sabbath
Her favorite pair.

And the heels go tapping:
With a clatter and a din,
From our old Vilna streets
They drive us to Berlin.

(David G. Roskies)

From MOTHER

1.
Friday evening coos up in the attic
and you tremble under the moon as it illumines
        your prayer book.
The tips of your yellow star pray
and shiver like parts of limbs.

The moon melts from your eyes,
the maternal drops brighten my faith.
Your prayer is like the warm *challah*
you feed to the doves.

In every one of your wrinkles I lie hidden,
I listen to you cough and shiver, but no one
should hear you breathe—since here, in a corner,
covered with earth, also lie my bones.

451

Your hand dozes on my forehead: be calm,
just another day or two and all will be well.
The other hand you place over my ear
so I should not hear the din of outrage.

4.

Where was *I*
when to the accompaniment of cymbals
you were dragged to the scaffold?
—My carcass was buried in a doghouse
with canine joy cursing itself.
On my lips—a leech.
In my ear—a spider.
I slunk out through a crack—
there beneath the moon that mirrors darkness
the wind played with pearls of snow.
And the snowy orchestra
swirling secretly
on the flag of the moon—
how does it display such beauty?
Each pearl plays
with its own shadow and the image
lightened my eyes
that I was prepared to howl thanks.

6.

I seek the four dear walls
where you breathed.

The steps below revolve dizzily
like a churning whirlpool.

I find my way to the latch
and tear open the door to you.

A tiny bird seems to weep
in the cage of fingers.

I enter the room
where your dream darkens—

The lamp you lit
feebly throws out some light.

On the table, a glass of tea
you barely managed to sip.

Fingers still move
on the silver rims.

The little tongue of light pleads
for mercy in the flickering lamp.

I offer my blood to the lamp
that it should not hold back its glow.

(Seymour Mayne)

J.L. TELLER (1912–1972)

A Galician by birth, Judah Leib Teller came to New York in 1921 and
studied at Rabbi Isaac Elchanan's Yeshivah, City College of New York,
and Columbia University. His volumes of Yiddish poetry include *Symbols,
Miniatures,* and *Poems of the Day.*

LINES TO A TREE

Around us speeches of birds. I tremble
with the passion before pairing.
We are both of one tribe.
My fingers dig in like roots
to the cool, dark depths.
For both of us the bark bursts,

the sap strives to emerge
and also the book.
The book and I, rooted like you,
with fear and trembling, involved
with rivers, earth and sun.
All nests are warm with whispering;
the birds tremble awake;
my hands on you, I seek your
knowledge, find your
longing.

(Gabriel Preil and Howard Schwartz)

MINOR KEY

The night lies blue and white
across the blanket.
My wife is porcelain
and I am like hail
in the windows.

Our dream repeats
our day.

You are right, comrades.
You smell of the revolution
like whiskey.
Your heads do not even turn.
When your stones hit my window
I won't care.

(Gabriel Preil and Howard Schwartz)

TO THE DIVINE NEIGHBOR

I turn to You
not at noon,
but at twilight,
when smoke writes in the air
and is at the same time
erased.

(Gabriel Preil and Howard Schwartz)

MIRIAM ULINOVER (1890-1944)

Ulinover was born in Lódz and died at Auschwitz. Much of her work, like
the two poems reproduced here, deals with motifs derived from Yiddish
folklore.

HAVDOLAH WINE

Everyone drinks Havdolah* wine
So I drink a few drops too.
Says Grandma sweetly earnest:
"My dear child: I must warn you

A girl who drinks Havdolah wine
Will grow a beard in no time,
So is it written in books
Over there in the bookcase."

I break into a cold sweat
And tap the edge of my chin:
Oh Thank God, still smooth and soft...
Only fright can make it bristle.

(Seth L. Wolitz)

IN THE COURTYARD

Summer—five o'clock
The courtyard's awake
Everyone's bustling about
Time to feed the hens.

Drawing the girl close to him
By the chicken coop
He gives her a tender kiss
And then a caress.

Frozen to the spot, the girl
Doesn't know what to do
And blood rushes to her face
The yellow hen gapes,

Should the hen peck a red dot
from her blushing eyes
Then all the egg yolks will bear
The dreaded blood spot.

(Seth L. Wolitz)

*Havdolah is the ceremony concluding the Sabbath.

455

MOSHE YUNGMAN (b. 1922)

Born in Eastern Galicia, Yungman was taken prisoner of war by the Russians in World War II. After the war he returned to Poland, and after a brief stay in Italy (where he led Zionist groups) he settled in Israel in 1947. His numerous books of poetry include *White Gates* (1964), *Smiles from the Holy Land* (1969), *Rainbows at the Head* (1973), and *In the Land of Elijah the Prophet* (1977). He lives in Kfar Tiv'on, outside Haifa.

MELONS

On small donkeys they bring in suns,
melon-gold which strains at the sacks,
The donkeys walk step by step
with slumbering chins
and drop the sound of bells note after note.

The street swallows the taste of melons.
And from barns, doors,
the donkeys are standing in pairs.
The entire world is suddenly sunny and bare
And it is small and good and warm to touch.

So drunk they walk about with melon suns.
The sewers overflow with gold.
And the mountains in the distance.
With small suns they are riding toward evening.

And suddenly nothing. The sacks are empty, extinguished.
The driver busies himself by his donkey,
and looks around for someone to come
to redeem him.

(Gabriel Preil and Howard Schwartz)

POLISH

ERNEST BRYLL (b. 1935)

Ernest Bryll is a prolific and popular poet, playwright, and fiction writer, one whose work often shows interest in formal control and in tradition. The poem here included comes from his collection *Adwent* (Advent, 1986).

IN A FEVER

I dreamed that in a light-industry mill
Where tired women choke and cough up dust
One worn-out woman failed her master's trust.
She couldn't stick it out to closing time—
It's rough but then the pay's not bad, and still
When the night shift's done you can stand in line,
Then get the kids to school with what they need—
Instead she broke down, and began to bleed.
Her white twill shirt turned red. Her piercing scream
Burst the factory gates, and in my dream
A horde of women poured into the street.

It overflowed, and no one dared to meet
Them face to face, though Warsaw phone calls warned
That this must stop . . .
                            The local party hacks
Sank clutching their receivers, drowned
In human seas. One of them yelled: "Hand out some meat,
And then remind these gals to watch their step,
Families can get hurt, better turn back
Now." Others whispered to the guards who kept
Watch over Them, the nation's true elite:
"Don't sweat, boys. It's just women. Running scared.
They're old and ugly, too. We couldn't care
Less . . ."

It was true. Their hair was gray with grime,
Their faces worn and wrinkled, and their weary
Bodies had paid the price of working overtime.
When they finally came in range, then you could see
That those who yelled at you, yelled toothlessly . . .

No, better wake up. Now. Before a terse
Voice gives the order no one can reverse.

6 April 1986

(Stanisław Barańczak and Clare Cavanagh)

ADAM ZAGAJEWSKI (b. 1945)

Adam Zagajewski, poet, fiction writer, and essayist, is, along with Ry-
szard Krynicki, the most prominent member of the "Generation of '68."
Since his 1972 debut he has published five more collections of verse.
His work has been translated into a number of languages. "On the Es-
calator" comes from *Tremor,* a selection published by Farrar, Straus and
Giroux.

ON THE ESCALATOR

How immovably they stand on the moving stairs,
those statues of my unknown
fellow men. Are you, too, among them?
How slowly they rise, without effort
and without fatigue. Down below, the city
which nobody will conquer anymore because
no one besieges walls nowadays. Destiny
surrenders gladly and these victors are no worse
than their predecessors. The sun
sets in the usual way, pink
cream touching the horizon. The streets
are open like empty beer cans, they sing
the same song, though not on command. The stairs
grow like pine forests. Why conquer cities,
hurl stones, burn temples;
laughter, a whisper, contempt are enough.
How impatiently the stairs grow, a pupa
turns into a butterfly. St. Bartholomew's

Night may last just five minutes, and without bloodshed,
only courage evaporates slowly. I look
at the crowd going up. So many faces,
so many cheeks, so many hopes, wishes,
clasped hands, in the irises of convex eyes
light crossing shadow. So many faces,
so many hands, and only one imagination. We
who ride down already know: no one is waiting
up there. Pigeons fight for food, swallows
write with quick hieroglyphs a letter
to the president, and the president laughs
like the wind.

(Renata Gorczynski)

STANISŁAW BARAŃCZAK (b. 1946)

Born in Poznań, Stanisław Barańczak attended and taught at Adam
Mickiewicz University; since 1981 he has taught Polish literature at Harvard. Besides his numerous books of poetry, his work has embraced
criticism, literary history, and translation into Polish of Shakespeare, the
English Metaphysical poets, and outstanding poets of the nineteenth and
twentieth centuries.

THE THREE MAGI

To Lech Dymarski

They will probably come just after the New Year.
As usual, early in the morning.
The forceps of the doorbell will pull you out by the head
from under the bedclothes; dazed as a newborn baby,
you'll open the door. The star of an ID
will flash before your eyes.
Three men. In one of them you'll recognize
with sheepish amazement (isn't this a small
world) your schoolmate of years ago.
Since that time he'll hardly have changed,
only grown a mustache,
perhaps gained a little weight.
They'll enter. The gold of their watches will glitter (isn't
this a gray dawn), the smoke from their cigarettes
will fill the room with a fragrance like incense.

All that's missing is myrrh, you'll think half-consciously—
while with your heel you're shoving under the couch the book they
    mustn't find—
what is this myrrh, anyway,
you'd have to finally look it up
someday. You'll come
with us, sir. You'll go
with them. Isn't this a white snow.
Isn't this a black Fiat.
Wasn't this a vast world.

(Stanisław Barańczak and Clare Cavanagh)

BRONISŁAW MAJ (b. 1953)

Bronisław Maj is, together with Jan Polkowski, the best representative
of the generation of poets that came into its own in the early 1980s. He
is the author of four books of poems, as well as editor-in-chief of the
journal *NaGłos*, published in Kraków.

May 13, 1981

The world: whole and indivisible, begins where
my hands end. As I stand at the window, I see it: the green spires
of Skałka and Wawel, the dome of St. Ann's, further, deep blue
hills, for so the woods look at dusk, beyond them
other valleys filled with cities, and still more cities:
on rivers, on wide plains sloping
to the sea, beyond which lies another sea, sharp brown
peaks, mountain passes, roads, and people's houses not unlike my
own. The breath that fills my mouth, lungs, blood is just
a share—mine only for a moment—of all the air
enveloping the world: indivisible. I see it—I know
that it is there, right at hand, at my fingertips, at my breath's warmth.
The rest is just a matter of miles, of imperfect vision—insignificant
on a scale of mind and heart. Hence right
at hand, just a few blocks off, on a large
city square full of people, my brother shoots
my father, here, at my fingertips.
Just like that: not a bang, not a whimper, like that.

(Stanisław Barańczak and Clare Cavanagh)

GERMAN

VOLKER BRAUN (b. 1939)

Born in Dresden, Braun worked in printing and as a machine operator in the mines before his studies in philosophy at the University of Leipzig. Since 1965 he has lived in Berlin, where he has worked with the Berliner Ensemble. Braun is a social critic, and is indebted to both the modern and the classical traditions in verse.

THE NEW PURPOSE OF HADRIAN'S ARMY

About the Emperor Hadrian
It has been reported from ancient times, between the lines,
That he didn't need his army in order to wage war—
For which purpose the empire was too vast, stretching
From Britain to Cappadocia or thereabouts—
But in order to travel.
Seeing that he could not disband the troops
In full view of the Goths or Sassanians,
He set them an unconventional goal:
Not to burn down cities, but to found new ones.
Wherever his spearmen landed, they reached for trowels
And whatever place they left had been rendered habitable.
He raised an obliterated Jerusalem from the ruins
And Athens, rich in palaces, he bedecked with palaces.
Before the countries on which he had set his heel
He bowed down . . . to study their mosaics.
All he took was the measurements of the devastated temples,
The cutting which he did was the famous "golden section."
Finally he gave orders to stone-masons, builders and smiths.
He chose his personnel, presumably, by reading in their features
Not warlike feelings but a feeling for art,
So that while they were still soldiers, they were already workers,
Still ready for battle but at the same time adept in art.
All this between the lines
In a language long dead. We in turn, however,
Think so forcibly along these lines
That this has for us the force of a parable,
That is to say, this was so long ago that it is almost true.

461

So can't we, in the upheavals we venture to bring about,
Use the whole state, while it is still necessary, like the army
For its own and for another purpose?
Being the master still of men, but at the same time
Making them masters of themselves, of their common sentiments?
This breach of convention which abolishes the rules
By becoming the rule?
This imperial luxury of the masses
Which makes the peoples rejoice
In their living languages?

(Edward Mackinnon and A. M. Elliott)

HELGA MARIE HEINZE (b. 1946)

Helga Marie Heinze, a native of Dresden, has published poetry, short stories, and socio-critical essays widely in distinguished periodicals. She is a contributing reviewer for *Die neue Gesellschaft*. The two poems here printed show this poet interested in verbal innovation in poetry.

LATE FLOWERS

captured septemberlight
red-blue suncontest
golden shrubchimes
autumn
spunsilky pompom
cutout blossomwool
sleepy-cool winterinkling
asters

(Katherine Bradley)

STORM

the horizon: washed ashore—
      swollen clouds;
the light: encircled—
      oatgold sultriness;

the city: crouched—
heavyhot stone.

blackflashing formations,
buzzing nosedives
on sulphurtarnished grain,
white position lights on their bellies;
—swallows—
heralds of the storm.

(Katherine Bradley)

CHRISTOPH EISENHUTH (b. 1949)

Born in Jena, Christoph Eisenhuth is a poet of secular and unorthodox
outlook, despite the fact that he is the author of liturgical texts. His
collection *Gespräche mit Christiane* (Conversations with Christiane),
published by the Union Verlag in 1977, is the source of the poem here
offered.

NIGHT IN THE VILLAGE

Dog-barking
Leaps over the darkness
Comes back from the hills
The night is an ear
For the voices of the yards

The black firs cup their hands
Around the lights in the village
The barn roof stoops
Under the weight of the stars

Dog-barking encircles the place
Guards the house
Does not pursue
The homeless man

(Allen R. Chappel)

THOMAS ERWIN (b. 1961)

Thomas Erwin, a native of Berlin, was not allowed to continue his education after completion of secondary school. He worked as a museum guard, was under arrest, and was permitted to leave for West Germany in February of 1981. *Der Tag will immer Morgen bleiben* (Day Always Wants to Stay Morning, 1981), the source of the poems below, appeared with the Piper Verlag in Munich.

PLAYING FOR KEEPS

Once
    I believed in exclamation marks
then
    I said "period" to that
now
    I love question marks
exclamation mark

(A. Leslie Willson)

FOR SOMEONE TEMPORIZING

You are wallowing in the catacombs of your anxiety,
carrying your shadow over puddles full of muck,
hoping still your image will not be blurred,
eternally you wait for great words, falling mute,
but I beg you, consider the question:
What approach-run is required to shoot past the goal?

(A. Leslie Willson)

LAND DONATION

Your disappointment gnaws holes
still there might be repairs
but Sister Needle repairs
by puncturing everything in all directions

it's also a trick of time
the wounds do stop hurting

but scars remain scars
only the skin gets wrinkles

(A. Leslie Willson)

CZECH AND SLOVAK

MILAN RICHTER (b. 1948)

Milan Richter was born in Bratislava, and holds a doctorate from the
Comenius University there. He has worked in publishing. Since 1981 he
has been a free-lance writer and translator. Richter often writes on the
Holocaust; he is also a Goethe scholar. His fifth poetry collection, *Ko-
rene vo vzduchu* (Roots in Air), appeared in 1992.

LIGHT?

> *Mehr Licht*
> —Goethe

More light
asked Goethe as he died.
(And his were not dark times.)
Since then we've had light in plenty:
The firing squads' bullets lit it
in the chest of the *Communards,*
like a fiery rose it flared
in the muddy trenches,
and on 10 May 1933 in Berlin's Opera Square
blond students fed it
with a deep black pile of hundreds of books.
(Behind the tall flames flashed the features
of a certain Dr. Goebbels,
but the dark times had begun already.)

Oh the light in the furnaces of crematoria,
the light of lamps on interrogators' desks,
the light over Nagasaki,
the light of assassinations, invasions and revolutions,
oh the light under the coats of titans and the SS!

More darkness
asks the poet, wishing to survive.
(And we don't have bright times today.)
More soft darkness in the wombs of cities,
darkness without alarms and the blood of television news,
darkness that shields you

from the electronic eye of police cameras
and spy satellites,
darkness without the decay of particles, cells, families . . .
Spotlit by a light which Goethe did not ask for
you long for the darkness of your own mind,
for two thousand years' oblivion . . .
and also for a little, a very little, good remembrance
which you would not owe to today's mephistos
haggling for the souls of those who're not for sale.

But *Mehr Licht*
is thought to have been Goethe's order
to his servant
to open the window wide.

(Ewald Osers)

ERIC GROCH (b. 1957)

Eric Groch, born in Košice, has two poetry collections: *Súkromné hodiny smutku* (Private Lessons in Sadness, 1989), and *Baba Jaga: Zalospevy* (The Witch Jaga: Mourning Songs, 1991). For the poem appearing below, he acts as his own translator.

FRIEND TO BIRDS

Old wigwam woman sitting beside the fire. And singing.
Who can't hear her?
A starved bird circling around her flesh.
Who is singing him?   •

We arose from the woman's cry and are trying
to fly up. Or at least light a fire.
Others are watching their star through the wigwam's eye.
But friend to birds is building a roof by
                                    turning up his head.
And he doesn't ask:
"Who lit the fire in the valley of the mother's pelvis?"
It simply burns: and the fire devours his questions
before they're uttered. Those are the ashes
he scatters on his head

with the romping of a child.

His face is an anti-reflection of the world.

His arms in the moment of uplifting
are the wings of some uncreated creature.

(the Author)

SYLVA FISCHEROVÁ (b. 1963)

Sylva Fischerová, a native of Prague, teaches philosophy at Charles University. She began publishing poetry at age eighteen. Her most recent collection of poems is *Velká zrcadla* (Large Mirrors, 1990), published by Československý Spisovatel. Fischerová's work shows great openness and vitality, as well as a sense of community and of inner form.

WHO KNOWS SOMETHING ABOUT WOMEN?

Above you, idols
hang grand and firm,
don't believe them, they're deserters
    from The Wheel of Fortune and never
    were beautiful;
now a girl in a dress
is waving at you,
in an ivory dress, her fingers
are copper
thimbles of the Fates,
        Add a log
        under Fate's kettle
        and warm your little finger;

boiled idols
fall into thimbles
and the girl has no face, she's a wild
western wind,
      ground trees down, towers, rocks,
      ate up the grass; some spurs
remained for those weeping,
left behind, bereaved; that's why
I lost you,

so the western wind
would swallow me, so boiled idols
would gulp down my gentle friends;
    who still knows
    nothing about women?

(Vera Orac, Stuart Friebert, and the Author)

MAGDA BARTOŠOVÁ (b. 1972)

Magda Bartošová is studying journalism in Prague. With the innovative, half-prose, half-verse, poem here printed, she is the youngest poet to appear in this anthology.

AND THAT IS WHY . . .

I keep losing.
I have already lost three umbrellas, two jerseys, a bag,
some abilities, a friend, money.
    Sometimes I find. A bill, a chestnut, earthworms in the
puddles, a boyfriend. And other marvelous things.

Mrs. Klucková told us don't be afraid of expressing your feelings by moving. Stand up at the street corner, tell people to go to hell, catch the falling snowflakes in your palms, dance with them, dance with the snowflakes over your heads and in your palms and in your eyes and mouths.

The ballet teacher Klucková had protruding collar-bones and that was what horrified and fascinated us at the same time.

We were all laughing at that notion—Mrs. Klucková dancing in front of the prefab. None of us, of course, has ever done it. We were nine and too sober and too proud.

Even now, the snow still falls, but we are not nine any more. And even if we dance from time to time in the mountains, far away from civilization, we are bashful of doing it in public.
    And that's why we're not worthy of winter.

(the Author)

470

HUNGARIAN

DEZSŐ TANDORI (b. 1938)

Dezső Tandori is a prolific poet, novelist (whose work in this genre includes science and detective fiction), essayist, and translator. In 1987 he was awarded the Attila József Prize. *Birds and Other Relations,* translated by Bruce Berlind, shows a connoisseur of birds (sparrows, with whom he shares his Budapest apartment) and of art.

UTRILLO: "LA BELLE GABRIELLE"

He was not left completely alone. Red.
There stood beside him, green, the good-hearted
innkeepers. Gray. Uncle Gay, the "Casse-
Croûte" 's owner, yellowish-violet, and Marie
Vizier, the proprietor of "La Belle
Gabrielle." Blue, black. They both
loved the eccentric, soberly decent and
childlike painter. Olive-green. When
he was alone, white, and didn't feel like
painting or drinking, sherry-color, he played
with a teddy bear and a toy railway. Prussian
blue. He spent most of his time in the back
room of one or the other of the two inns. Brown.
He didn't go out to paint; azure and orange;
he knew if he went into the street, sooner or later
he'd have one too many, pink, would pick a quarrel
with the passersby, greenish-white, and at such times
it was always he who came in second. Black. He
bought picture postcards and carefully enlarged them. Yellow, reddish-purple, green.
Since, white, olive-brown, he hadn't received a thorough
training, he couldn't paint "from memory," carmine;
he needed the sight or a substitute for
sight, orange—the picture postcard. The
colors, however—red, gray, green,
black, white, brown, yellow, orange-red,
reddish-purple, and so on—he himself superimposed
on the gray, white, blue, carmine photograph.

(Bruce Berlind)

GYÖRGY PETRI (b. 1943)

György Petri, a major dissident voice, studied philosophy at the University of Budapest; since 1974 he has been a free-lance writer. His collections of poetry include *Magyarázatok M. számára* (Explanations for M., 1971), *Körülírt zuhanás* (Plunging Circumscribed, 1974), *Örökhétfő* (Eternal Monday, 1981), *Hólabda a kézben* (Snowball in Hand, 1984), and *Azt hiszik* (So They Believe, 1985).

ONCE AGAIN

Once again
the solitary swimming pools.
The beach chairs.
Lavish sunshine
spraying tired bodies.
Soon the wind
will peck away
at my body-long watery imprint.
The shower is a stupor
of heat and vapor.
Booming on the nape of my neck:
           a fragmented ray.

(Robert Austerlitz)

XENIA

Light, water, salt, bread,
dill and mustard seed:

Our cucumber has gingerly
matured into a pickle.

It has absorbed the elements.
The child of nature and of art.

(Robert Austerlitz)

ZSUZSA RAKOVSZKY (b. 1950)

Zsuzsa Rakovszky was born in Sopron and obtained her certification for teaching Hungarian and English from the University of Budapest. Between 1975 and 1981 she worked as a librarian; since 1987 she has been free-lance. For her first two poetry collections: *Jóslatok és határidők* (Prophecies and Deadlines, 1981) and *Tovább egy házzal* (One House Up, 1987), she won the Attila József Prize in 1987.

DIFFICULTIES OF FALLING ASLEEP

The day ground down; but sheer walls
waver, turmoil: shining spokes of wheels
are thrown by a chandelier onto the ceiling
where the glass folds and bends; in a circle of light
beetles swarm—a wreath of dots—
or blaze in large shadows across fields of light.

Windowpane: in gray waters a room of coal.
The lamp's twin sways, deprived of its glow,
flat black shapes hover, dim—a corpse's
remnant of memory. My living
flesh, in its sweat-soaked nylon, prickles under the light,
melts it, bedazzles: silken sugar candy.

The clock face makes light: I don't dare sleep.
In the chambers of my dream the deceased awaits me:
crouching on the sofa in her usual place,
or ready to leave: adjusting her knitted
skirt, the lining rustles, lapping against
her tights. She paints her lips, her eyebrows,

or looks for book or basket of knitting needles.
I evade her glance: she must not know . . .
by daybreak she'll be paper, whirled by the wind, empty rags.
Again and again she comes, whenever I drift
into interrupted dream, from which once more
lightning wakes me: leaning against a radiant sky
the blinding, zigzag profile at my window.

(Barbara Howes and Margot Archer)

FLÓRA IMRE (b. 1961)

Flóra Imre, born in Budapest, obtained her teaching diploma in Hungarian, Latin, and Greek from the University of Budapest in 1985; since then, she has taught at the József Eötvös Gymnasium there. Her poetry collection *Az Akropoliszra néző terasz* (Terrace Looking Out on the Acropolis, 1986) reflects the interest in Greece that the poem here offered also shows.

SO, THEN, THE SOUL

So, then, the soul craves an earthly body,
One that can touch, one you can touch in turn.
Thera: a scraggy place; time, thick as honey,
Blends its fever-wracked trees, makes them merge.

Now it's with all but unbearable tension
That the forms crowd, jostle in the eye.
A cubic shape: a stone, spotted, brown.

The soul would look; nothing yet it can see.

And then at once, unrestrained, this mute,
Brimful, objective state comes on flooding,
And it sees trees, dried-out, crooked trees,
Yellowish rocks, sunbaked houses'
Sharp-edged white; and, as the donkeys
Stop near the wall, it sees their bodies'
Bluish, grotesque shadows in the bright
Dust, bellying, as the noon sun shines.

And it sees; somehow in a trance,
Objects suddenly turn gleaming bright.
Spaces open; spacious, clean lines,
Loosened into distance, fresh, uncovered.

And when it has just attained to touch,
All but reached the downy, desired shell,
Eyesight at once leaves it in the lurch.

It craves an earthly, an earthly body, the soul.

(Emery George)

474

ROMANIAN

ANGHEL DUMBRĂVEANU (b. 1933)

Anghel Dumbrăveanu was born in Dobroteasa, in the southern Romanian region of Oltenia. His over thirty volumes include fourteen of poetry, three novels, and six volumes of translations. Dumbrăveanu is an important member of the generation of the 1930s and early 1940s, an original whose lyricism is born of precise observation, imagination, and feeling.

GENESIS

The world was made by an Egyptian god
On a potter's wheel. He took earth and water from the Nile
And spun them round his thought into light.
Out of light he conceived woman and flowers.
But woman looked like a pitcher, and when he raised her to his lips to
    drink,
The potter grew drunk—his hands
Wildly caressed the pitcher upturned beside him.
Then song was born . . .

(Adam J. Sorkin and Irina Grigorescu Pană)

SEEDS OF FLIGHT

Summer in the Argeş, with plum trees bent low on the hills.
A child is carving whistles out of willow shoots. How far away
Are all these things. For a season, the vineyard thrush
Will flit along our roads choked with wild barley
And warble blue tears. The leaves grow rusty
On the walnut tree under which we'll never return.
In vain the horses are waiting for us near the river. Spiders
Dangle from the stars on yellow ropes and lift themselves high up
Under the dome of crickets. The elements
Lose their substance, to be born again suspended
In the glass globe of the world. How far away
Are all these things. The thrush will not find the leaf
Of your hand, with the seeds of flight. And to whom
Will it sing a lament of the lost road that once was ours?

(Adam J. Sorkin and Irina Grigorescu Pană)

GRETE TARTLER (b. 1948)

Born in Bucharest, Grete Tartler is of German background and a major talent on the present-day Romanian poetry scene. She is also an accomplished translator from the English and from the Arabic. Her collections of poetry, one of which won the Poetry Prize of the Writers' Union, include *Substituiri* (Substitutions, 1983) and *Achene Zburatoare* (Winged Seeds, 1986).

SIGHIŞOARA 1982

If you had the insight
as the river has, flowing between
yellow and green houses with chained doors,
breathing out two breaths—the scent
of lime trees and that of damp—
listening to the organ music of tiles:
   piano (the ones with moss on them)
   forte (the redder ones).
If you had the impulse
as the cannon do, booming
   over ANDREAS RATH, IRONMONGER,
   over the high school, over the volley-ball field,
   over the jasmine and gnomes in the garden.
If you would bring supplies
from the other side of the river!

To be clouded and indifferent as silt!
You have lived many lives here in vain.
The water carves a scar
on the face of the town;
even you have set the gold ring
from your finger under your eyelids now:
a snake painted on the chemist's shop wall—
you read there "Water Level, 1970."
But you, so indifferent, don't bring
supplies, although you've been crossing
to and fro across the river for so long.

(Fleur Adcock)

476

DANIELA CRĂSNARU (b. 1950)

Daniela Crăsnaru was born in Craiova, studied Romanian and English at Craiova University, and has worked in publishing. She has to her credit twelve verse collections, in addition to children's books. Her recent collections include *Niagara de Plumb* (Leaden Niagara, 1984) and *Emisferele de Magdeburg* (The Hemispheres of Magdeburg, 1987).

THE LIONS OF BABYLON

A morning
five thousand years ago.
In the stone cellars
of the palace,
to the left and right of the scribe,
the lions of Babylon.
He
has never seen the sun,
he hasn't seen the river,
hasn't seen the sea.
He is now bent over, writing
on a clay tablet
about the sun.
He gives a minutely detailed description
of how the great river flows
into the sea.
The slave who is dictating to him
has for his part also never
seen the river,
the sun, the sea. He too
has heard of them—
from another slave
who glued his whole body against a wall
until he lost his breath,
until he bled,
until his body caught
the light from outside,
the murmur of the river as it flowed
into the sea.

(Fleur Adcock)

ION CRISTOFOR (b. 1952)

Ion Cristofor was born in the village of Geaca in the county of Cluj, and is a graduate of the Faculty of Philology, Cluj University. Since 1973 he has worked on the staffs of literary magazines at Cluj. Cristofor is the author of two collections of poetry: *In Odaile Fulgerului* (In Rooms of Lightning, 1982), and *Cina pe Mare* (Supper on the Sea, 1988).

NIGHT TRAVELLER

Your earth colored cheeks
with their lines recalling
the sun's writing on endless headstones
and the houses inhabited by the dusty south wind
by a gypsy woman looking after cats
whose mewing opens the locks of night.

The tramp of the barbarians, the tide of cries and shadows
flooding the golden pavilion in which
blind memory walks
through a labyrinth of mirrors.
The wedding chamber of the beast resounds with the song of
                                       the barrel organ
reptiles with phosphorescent blood
leap out from under the phrases of the scribe.

The sun returns to the trees of night
a cathedral on fire floats over waves
moved by your breeze
underground choirs appear from under the threshold
of the house

The lioness near a flame licks her cubs
ignoring the wind which builds on dunes and cliffs
the sand kingdom
of your disquiet.

(Brenda Walker and Michaela Celea-Leach)

YUGOSLAV

NOVICA TADIĆ (b. 1949)

Novica Tadić was born in Montenegro and lives in Belgrade. He is a highly original poet; his seven collections of poetry often concentrate on a world of surrealism and evil. "Thief" and "At the Hairdresser" come from *The Horse Has Six Legs: An Anthology of Serbian Poetry*, edited and translated by Charles Simic.

THIEF

Drags his tail on the ground, damn thief
Feeds his thousand mice
Eats, drinks up, guzzles
All day long sings the blues
Parties like no one
Shears with happy ears
Drags everything down down
Trades truly with hollow gold coins
Steals and filches
Leaves in testament walls, ashes, air
Stuffs his gravelike sack

Crawls into the egg of a snake

Runs backwards toward chaos

(Charles Simic)

AT THE HAIRDRESSER
(phantasmagoria)

At night at the hairdresser's
The angel with bright scissors
And a monstrous comb draws near
To the archangel's funnel-like ear:

If God is dead, if he truly fell in
The abyss, let's place instantly
On his empty throne, the hairdresser
Who does our hair so well.

(Charles Simic)

LILJANA DIRJAN (b. 1953)

Liljana Dirjan was born in Skopje; she works for a woman's magazine.
The author of several volumes of poetry, she has won two prizes for her
work.

POETRY EVENING

We sat among Emily Dickinson's 1775 poems
(that, at least, was Thomas Johnson's approximate sum)
those who knew English and seized upon finesses
they were passing a pencil to each other to mark the powerful passages
They talked about internal rhymes
untranslatable structures
complimented each other on their skill and precision
every so often jumped up from their seats
and returned to the text again

Between them, like a second, I waited:
which would draw his gun to kill me?
Poor Emily
could never have expected
that her poems would be some use
that they would kiss in the air . . .
While I crumpled up
the target grew
and they fired like poetic experts
accurately
without noticing in their excitement
how I unbuttoned my life
and faded away.

(Ewald Osers)

NINA ŽIVANČEVIĆ (b. 1957)

Živančević, a native of Belgrade, is a journalist, literary critic, and translator, as well as a poet. For several years she lived in New York, writing in English, and now lives in Paris. "A Poem with a Tilde in the Title" comes from her poetry collection *Duh renesanse* (The Spirit of Renaissance), published in Belgrade in 1989.

A POEM WITH A TILDE IN THE TITLE

I'm sad and serious
like the little Donna Infanta imprisoned
in Velázquez's painting, while I watch
my silly royal retinue with remote
and soulful eyes, bending over a ray of light,
directed at young princes,
niños, and dueñas, and while I shrug
the pearls of dust off my lonely shoulders,
and hide in my curls their friendly deceits,
and wait for my dwarfs and polite servants
to bring me the indifferent cup of chocolate . . .

I'm sad and serious as if yesterday
I sent a fleet of explorers
to search for a new continent and bring me back fresh spices,
golden idols, new refinements,
and the exaggerated insults of distant monarchs,
soft fabrics and unusual toys in the shape
of the bleeding human heart, while knowing
that all they will bring me back
are barbaric seashells, effigies of gods
with the hollow, painted mouths
who are trying to say something simple and terrifying
that is like a whisper, a bottomless howl . . .

(Charles Simic)

SLAVE GORGO DIMOSKI (b. 1959)

Slave Gorgo Dimoski was born in a village near Ohrid, Macedonia. He is employed as a schoolmaster. He has to his credit several volumes of

481

poetry, and is the winner of two prizes for young authors. His work is characterized by the lyric concision and silence of which "Revenge" and "Rest" are two good examples.

REVENGE

The executioner arrives on the scene
and carefully raises his axe

that is his duty

      Out of turn
      leaning
      without beginning

through wellspring, future and light
flashes his sword—even then

in the dark dawn the miraculous dance
with the executioner is pure.

(Ewald Osers)

REST

It's time to sit down
on the trodden butterflies
of our dream

To breathe within ourselves
to enter erect

A bird flies over us
a pensive billy-goat stands next to us
the wind strikes our heads
it certainly is a good day

in the upturned cup
above us.

(Ewald Osers)

482

BULGARIAN

MIRYANA BASHEVA (b. 1947)

Miryana Basheva is an outstanding woman poet of the generation of the 1940s, and a recognized spokesperson of young people. Born in Sofia, she studied English language and literature at Sofia University. She is a poet of the modern city and of alienation, and frequently uses the irony and detachment evident in "Business," the example of her work offered here.

BUSINESS

He comes often. On business.
Honest, hard working . . . That's all.
But someone is head over heels in love!
And I'm head over heels in pain . . .

He comes mostly in the evening.
Normally sitting in the corner.
And I feel like game
and my gaze is rounded.

He's not there for personal reasons! . . .
He leaves pure as an angel . . .
Clicks the lock monstrously,
as if he's just fettered me.

Like a jailer, he deliberately
strides the stairs.
And I hide from freedom
in a jailhouse of dumbness.

Secretly I sharpen my sword,
my teeth and guillotine blades.
There's more business to be done!
He's bound to return—
after all, what choice does he have . . .

(Brenda Walker, Belin Tonchev, and Svetoslav Piperov)

GEORGI BORISOV (b. 1950)

Georgi Borisov studied at the Maxim Gorky Institute for Writers in Moscow, and is a highly regarded translator of Russian poetry. He is one of the chief editors of the Bulgarian periodical *Fakel* (Torch). Borisov's work shows formal control and dramatic power, both clearly indebted to his Russian studies.

UNTITLED

And when the winter wind rushed straight into my heart
and the wolves' centuries-old howl resounded in my throat,
and the night peered at me, her glance bristling and mad,
I got up and cursed the night, my blood, the sun.

And I slipped noiselessly through the peaceful city like a beast,
beside hotels and hospitals, taverns, cars and shops,
and I saw the last kerosene drops blooming
like blue poppies on the wet, black asphalt,

dead neon moons suspended in street corners,
the last lovers walking in the dark gardens,
and I heard the maternity ward echoing with songs,
the last trains whistling their sharp farewells.

And when I reached the middle of the deaf, black field
and the earth's salty blood licked my palms,
and when overhead the last snow began to fall
and heaped the night, the furrows and the muddy road,

the wind carefully swept my tracks,
its swirls wove an icy noose around my throat,
and the earth was hanged, and my body swayed
like a last song under the bright, sharp stars.

(Lisa Sapinkopf and Georgi Belev)

LYUBOMIR NIKOLOV (b. 1954)

Lyubomir Nikolov is a respected poet and translator. He recently participated in the International Writers' Forum at Iowa City, and in January of 1990 he was chosen Poet of the Month by BBC Radio Three. His two volumes of poetry are *Called by the High Tide* (1981), and *Traveller* (1987).

AXES

The axes lie next to the river.
And their smooth handles glisten.
On many stumps the black ants swarm.
Like African housewives they bustle.

Birds fly above the branches.
In the branches themselves sap rises.
The axes lie waiting and near.
The axes lie next to the river.

O, how we honed them once! Against
our thumbs we tested them all summer.
And looked at ourselves in the blade.
The axes lie next to the river.

It smells here of lichen. Frogs croak.
Deep in the valley fog clears,
With rakiya we rinse out our throats.
The axes lie next to the river.

The axe-heads—stainlessly cool
and dead cool the polished handles
lie. They wait for us to seize them.

And they rust in the spiky grass.

(Roland Flint)

IVAN KRUSTEV (b. 1965)

Ivan Krustev is at present pursuing graduate studies at Kliment Ohridski University, Sofia. His first poetry collection is entitled *Reading in the*

Dark. Krustev is one of the founders of the samizdat press in Bulgaria, and is at present a member of the Bulgarian Independent Literary Society.

NOSTALGIA FOR TITIAN

Her breasts are the cities,
they are the view which opens
from pons Veneris and from the book
overgrown with Renaissance nettles.

The fresco wriggles like a snake,
humidity mines it like a mole.
In the library, in the seventh row,
the Madonna learns to speak.

The canals of the one-time Republic
bear a strange analogy:
Is the city a replica of the fresco,
or does my face look like Venice?

Does my face, dug by words,
conform to the idea of eternity?
To the left of the chin's square
the market road starts.

A hundred steps to the right,
and I kneel before raped Europe.
Fishermen long ago have set
the nets of allegory.

They only have to drag the draught out
and feed the library.
The fish bellies are shiny
like the breasts of the Madonna.

Her breasts are like cones
driven into the infant's body.
I hear screams, splashes and songs . . .
The Madonna is a sand-glass.

(Belin Tonchev)

YIDDISH

H. LEYVIK (1888–1962)

H. Leyvik (pseudonym of Leyvik Halpern) was born in Ihumen, a small town in Belarus, and received a traditional Jewish education. After imprisonment and exile in Czarist Russia, he came to the United States in 1913. In 1958 Leyvik received an honorary doctorate from Hebrew Union College. "Open Up, Gate" commemorates his stay at a TB sanatorium in Denver between 1932 and 1936.

OPEN UP, GATE

Open up, gate,
Threshold, you tell—
I am coming again
To an intimate cell.

My body—fire,
My head—snow;
And on my shoulders
A bag of woe.

Farewell. Farewell.
Hands. Eyes. Bowed.
A goodbye on the lips
Flared—burnt out.

From whom did I part?
Farewell to what past?—
Perennial questions
This time don't ask.

In fire, in flame
The prairie spreads,
And snow in the glare
Of mountainous heads.

I bring to your feet
My bag of woe,

Land Colorado
Of fire and snow.

(Benjamin and Barbara Harshav)

MELECH RAVITCH (1893–1976)

Melech Ravitch (pseudonym of Zekharye Khone Bergner), born in East
Galicia, was active as a writer in Vienna and in Warsaw, and settled in
Montreal in 1941. His work includes many volumes of poetry, essays,
and memoirs. His collected poems *The Song of My Songs, 1909–1954*
(1954) and the subsequent *Post Scriptum* (1969) drew on thirteen earlier
collections.

VERSES WRITTEN ON SAND

In the garden. Summer's end. Evening. On a bench.
The second highest tower burns in the west.
The night wind has risen and with the rake
I begin to compose a poem in the sand:

We are destructive, and friends' blood
is as thin as water to us.
Can I ask God or man why this is so?
It seems relatively easy to be good.

Like the slaughterer's knife
we are always in the right.
I ask you again, God or man.
It seems so difficult to fully give way to strife.

We destroy, but who else sings on
about turning the other cheek to the oppressor—
And under our jackets we can hardly hide
the newly acquired weapon.

In the garden. Summer's end. Evening. I rise from the bench.
The darkness has eclipsed the last of the towers.
I simply say farewell to the emptiness
and in the darkness trample on my poem written in the sand.

(Seymour Mayne and Rivka Augenfeld)

MALKA HEIFETZ-TUSSMAN (1896–1987)

Malka Heifetz-Tussman, born in a village in Volyn, Russia, came to America in 1912. Tussman studied at the University of Wisconsin, and was literarily active in Chicago, Los Angeles, and Berkeley. Her collections of poetry include *Am I Also You?* (1977) and the posthumous *With Teeth in the Earth* (1992). She was awarded the Itzik Manger Prize for Yiddish Poetry in Tel Aviv in 1981.

THUNDER MY BROTHER

Thunder my brother,
My powerful brother,
Stones rolling on stones—your voice.
Like a forest, forceful, your voice.
What pleasure you take in making mountains rattle,
How happy you feel
When you bewilder creeping creatures in the valley.

Once
Long ago
The storm—my father—
Rode on a dark cloud,
And stared at the other side of the Order-of-the-Universe,
Across to the chaos.
I, too,
Have a voice—
A voice of fearsome roaring
In the grip of my muteness.

And there are commandments
Forbidding me:
"Thou shalt not,
Thou shalt not"
O thunder,
My wild unbridled brother.

(Kathryn Hellerstein)

BERYSH VAYNSHTEYN (1905–1967)

Berysh Vaynshteyn was born in Reyshe (Rzeszow), Galicia. At age eighteen he went to Vienna, and two years later he settled in New York. Vaynshteyn's trilogy of epic, book-length poems encompasses his three homelands: *Reyshe* (1947), *America* (1955), and *King David's Estates* (1960). "Mangin Street" shows a poet strongly indebted to Whitman and to Sandburg.

MANGIN STREET

Here too there are guys like the guys of the Volye.
Though in their clothes they look decent, clean and slender,
They have gentle faces and hands with soft nice fingers
That want to play with amber pistols and not with carcasses.

Guys of Mangin Street have a heart that lusts to kill a man;
They're always talking of murder-blood and know how to outsmart
    strict uniforms.
Peeling pictures from Italy still hang in their dark houses.
Alien to them is their mother's eternal sorrow, their father's language
    from home.

They are sons of fathers who plod in the mornings to the docks with
    thick-veined hands
Or stand wet in dug-out pits and throw up shovels of deep earth.
Tired as their fathers, the sons fall ragged in their clothes on the bed in
    a New York dawn;
And from their sleep blows a foamy sour smell of embraced women.

When twilight covers the panes of their houses, the mothers go out to
    the street
And overturn rich garbage bins where cats feed cleverly in the refuse.
Their daughters have a weakness for silk and for guys like their broth-
    ers
Who can see the blood of a murderer's wound run drunkenly in a New
    York dawn.

(Benjamin and Barbara Harshav)